ADVANCE PRAISE FOR

Is Ultimate Reality Unlimited Love?

"It takes chutzpah to attempt a book like this. We are such partially seeing creatures, barely glimpsing splinters of truth, savoring here and there tiny tastes of human love, and, in moments of meditation, perhaps getting a peek into the Divine Heart of Love—everything melting and moving, flowing, ever changing, ever eluding our grasp. So, hats off to you, Stephen, for helping us to better understand Sir John Templeton's hopes for us to grasp Ultimate Reality. This isn't simply a book to read. This is a deep ocean luring us to plunge in."

—SISTER HELEN PREJEAN, author of *Dead Man Walking*

"In this beautiful homage to Sir John Templeton, who put Unlimited Love at the center of his life, Stephen Post reminds us that *agape*—universal, unconditional love—is an essential quality of the most profound understanding of Ultimate Reality. Just as when there is a sun and light rays will effortlessly shine on all without partiality, the realization of Ultimate Reality will spontaneously express itself as Unlimited Love. In showing this in multiple ways, Post's work is a precious gift to humanity."

—MATTHIEU RICARD, Buddhist monk and scientist, recipient of the French Order of Merit, founder of Karuna-Shechen, and coauthor of *The Quantum and the Lotus*

"What is it to be a material success? By any standard, Sir John Templeton embodied the answer. But he also epitomized metaphysical success. In this unprecedented book, Stephen Post draws on the private words and personal practices of a business icon whose virtues transcend time. What is it to leave a legacy in more ways than one? Read on."

—PROFESSOR IRSHAD MANJI, *New York Times* best-selling author of *Allah, Liberty and Love* and founder, Moral Courage Project

"Dr. Howard Thurman said: 'We are eternally obligated to grapple with great ideas.' Stephen Post is superbly equipped to grapple with one of Sir John Templeton's great ideas. Through this timely publication we can all participate in this creative dialogue."

—OTIS MOSS JR., DMIN; pastor emeritus, Olivet Institutional Baptist Church; internationally renowned civil rights advocate; advisor to Presidents Carter, Clinton, and Obama; chairman emeritus of the Morehouse College Board of Trustees

"As one who worked closely with Sir John Templeton and had the precious opportunity to be engaged with his ideas, this thoughtfully written book by Stephen Post captures an essential aspect of Sir John's thoughts and hopes with regard to Ultimate Reality and Unlimited Love. Sir John's deep unselfish love for humanity, his bold metaphysical ideas, his humility as a magnificent soul, and his nobility of spirit echoes throughout the pages."

—REV. REBEKAH ALEZANDER DUNLAP, minister and author, Unity School of Christianity

"As an act of love and loyalty, Stephen Post helps us understand the mind and spirit of Sir John Templeton. A fierce and passionate man, Sir John, as Post affectionately calls him, had the courage to ask the big questions and the tenacity to seek answers. So is Unlimited Love the Ultimate Reality? Read this with a curious mind and an open heart and you will find yourself experiencing something divine. Love."

—DANIEL GOTTLIEB, PHD, host of the radio show "Voices in the Family" and author of the best-selling *Letters to Sam* and of the forthcoming *The Wisdom We're Born With: Restoring Our Faith in Ourselves*

"Sir John is very well situated in this book with reference to Unity and New Thought."

—REV. GLENN R. MOSLEY, PHD; president-CEO emeritus, Association of Unity Churches International; author of *New Thought, Ancient Wisdom*

"This is a fascinating account of a very American businessman and his profoundly universal vision of love as the Ultimate Reality."

—WILLIAM C. CHITTICK, PHD, author of *Divine Love: Islamic Literature and the Path to God*

"With the relentless focus on extremism, many regard the global resurgence of religion cynically as the 'Revenge of the Sacred.' This courageous, important, and timely book proclaims a contrary view of the Sacred as Ultimate Reality and Unlimited Love, and its catalyst is the vision and life of Sir John Templeton. Stephen Post deftly educates cynics about the perennial and profound idea of Divine Love at the core of all religions, now obscured by a materialistic modernity dominated by narrow scientisms and religious arrogance. Together, Post and Templeton inspire us to recover and reconnect with that core in our different traditions. Outside America and the Christian West, ideas about the Sacred as Ultimate Reality and Unlimited Love would overwhelmingly *not* be regarded as unusual. If anything, with the steady increase of religious extremisms everywhere, this book will be music to the ears of millions of Hindus, Muslims, Buddhists, and those who are intellectually and mystically inclined. Templeton may have winced in humility, but it is likely that many would regard him as a guru, sufi, teacher, sage, and, at the very least, the 'Best of the West.'"

—DR. DURRE S. AHMED, chairperson and senior research fellow, Center for the Study of Gender and Culture, Lahore, Pakistan

"In this book devoted to an exegesis of Sir John Templeton's thoughts on the prime role of love in the universe, Stephen Post lovingly weaves a reminiscence of his mentor into a compelling narrative filled with erudite associations that may be provocative to the modern ear but that are also illuminating. If Erasmus was right and 'humility is truth' perhaps we shouldn't be so fast to foreclose on the possibility that love is the ultimate hidden variable."

—DR. GREGORY L. FRICCHIONE, MD, director of the Benson-Henry Institute for Mind Body Medicine at Massachusetts General Hospital and professor of psychiatry at Harvard Medical School

"I love all of Stephen Post's memories and anecdotes of Sir John Templeton. I met him just once, late in his life, sometime in the 1990s when he was in London to give prizes in the essay competition among school children, and I remember vividly the radiance of his face and his kindly disposition. This extraordinary book presents Sir John's life-guiding vision of the ardent symbiosis between Unlimited Love and Ultimate Reality. Based on his personal knowledge of Sir John, his numerous writings, and correspondence, Post has done a marvelous job in conveying the strength and boldness of a vision that can change humanity and create a better future for all."

—URSULA KING, professor emerita of theology and religious studies, University of Bristol; professorial research associate, Department of the Study of Religions, School of Oriental and African Studies, University of London.

"This book is written as an almost sacred obligation to honor the dying request of Sir John Templeton—that the book he wished he could have written would be written. It has and it surely is as close to the vision Sir John would have written in response to his bold question, 'Is Ultimate Reality Unlimited Love?' Stephen Post has done a remarkable job in being the faithful voice of this 'last book' of what I fondly think of as the 'Mystic of Winchester.' Here is the vision of a man whose immense wealth has always been in the service of seeking an answer to what is perhaps the greatest question that can be asked."

—RALPH W. HOOD JR., former editor of the *Journal for the Scientific Study of Religion* and recipient of the William James Award from the Division of Psychology of Religion of the American Psychological Association

"The mystical and Eastern dimensions of trailblazing philanthropist Sir John Templeton's worldview are far too often overlooked. Stephen Post has balanced the record in a scholarly tour de force. Drawing on many unpublished letters and other sources, Post traces influences from Unity spirituality, the perennial philosophy, and Emersonian transcendentalism, setting them alongside Sir John's better-known Presbyterian influences. *Is Ultimate Reality Unlimited Love?* offers the fullest integrated view to date of Sir John the man and his mature vision, a vision that continues to live and influence the world through his foundation, writings, and legacy of personal inspiration."

—DOUG OMAN, PhD, School of Public Health, University of California at Berkeley

"Sir John Templeton's remarkable investment acuity is paralleled by his dedication to the scientific understanding of the power of love and to the dissemination of that knowledge. Thank you, Sir John."

—HERBERT BENSON, MD, director emeritus, Benson-Henry Institute for Mind Body Medicine at Massachusetts General Hospital; Mind Body Professor of Medicine, Harvard Medical School; author of *The Relaxation Response*

"Humility, purpose, and love—Unlimited Divine Love—these were the big ideas that animated Sir John Templeton, and which his 'surrogate writer,' Stephen G. Post, faithfully conveys on his behalf. By blending intellectual biography with Sir John's spiritual worldview, Post offers a helpful and hopeful alternative to today's 'me' generation."

—DAVID G. MYERS, PhD, professor of psychology, Hope College

"This is a beautifully written testament to Sir John Templeton in one of the key areas that he was passionate about. It is written by someone who knew Sir John well and who has spearheaded efforts to make Sir John's dream of spreading the message of unconditional love a reality in a world that desperately needs it. This is a 'must read' for anyone wishing to do the same."

—HAROLD KOENIG, MD, director of the Center for Spirituality, Theology, and Health, Duke University

"Part intellectual biography, part Festschrift, part spiritual treatise, Stephen Post has ably captured Sir John Templeton's remarkable vision of

Unlimited Love as the source and foundation of all being. The world needs this redemptive and transformative message now more than ever."
—JEFF LEVIN, PhD, MPH, University Professor of Epidemiology and Population Health, Baylor University; editor of *Divine Love: Perspectives from the World's Religious Traditions*

"Stephen Post captures the essence of the life of one of our truly great spiritual leaders, Sir John Templeton. Post eloquently elucidates Sir John's belief that universal love offers the possibility of transforming our focus on material wealth to one of elevating the importance of the spiritual roots that enrich and give meaning to our lives—a critical message for our times."
—CLAIRE HAAGA ALTMAN, president and CEO of Volunteers of America–Greater New York

"Stephen Post tackles a number of challenges in *Is Unlimited Reality Unlimited Love?* First, writing about the importance of unselfish love in a well-lived life during an age of greed and narcissism, and second, to undertake this discourse in the voice and thoughts of Sir John Templeton is courageous—some might say lunacy. At a time when so many young people are seeking meaning in their lives, I can think of no more important subject to address and no more qualified individual to address it than Post."
—STAN ALTMAN, PhD, professor, School of Public Affairs, and former dean and interim president, Baruch College, CUNY

Is Ultimate Reality Unlimited Love?

SIR JOHN TEMPLETON AND STEPHEN G. POST

Is Ultimate Reality Unlimited Love?

In Humble Response to a Request Made by
Sir John Marks Templeton (1912–2008) in his Last Days
that a Book be Written to Faithfully Consolidate
his Thoughts on His Quintessential Question
Using a Title He Designated

Stephen G. Post
for/with Sir John Templeton

WITH A FOREWORD BY
Drs. John M. and Josephine Templeton

TEMPLETON PRESS

Templeton Press
300 Conshohocken State Road, Suite 500
West Conshohocken, PA 19428
www.templetonpress.org

ISBN-13: 978-1-59947-451-9

Designed and typeset by Gopa & Ted2, Inc.

Library of Congress Cataloging-in-Publication Data on file.

Printed in the United States of America

14 15 16 17 18   10 9 8 7 6 5 4 3 2 1

Dedicated to Dr. John M. Templeton Jr., MD &
Dr. Josephine "Pina" Templeton, MD

Contents

Foreword

THIS BOOK is extraordinary and very inclusive. It conveys admiration for the mind of Sir John Templeton, for his key formative influences, and for the scientific rigor that Sir John emphasized in his quest to seek and anticipate at least increments of the nature of "Ultimate Reality." Author Stephen G. Post took on the role of a humble student over many years so that he could delve into the essence of Sir John's writing and thinking, and like Sir John himself, "probe questions" with the application of the logic of rationality.

The book is divided into three parts which should be read in sequence to fully understand the thought, motivations and progressions of Sir John's mind. However, the reader can also choose to read selected parts that he or she is most drawn to at any moment.

In Part One, Dr. Post makes the case for the essential premises that are so central to the overall book, all of them grounded directly in Sir John's own vision and cognitive framework, his major intellectual influences, exegesis of his writings on Ultimate Reality and Unlimited Love, and from various letters of correspondence from Sir John to Dr. Post. We see clearly Sir John's continued and persistent open-mindedness with regard to this tremendous metaphysical and scientific question, *Is Ultimate Reality Unlimited Love?* Sir John was always willing to ask the biggest questions, and this one was at the core of his mind and being consistently over the course of his lifetime, and as his life was ending.

Sir John could not consolidate all his thinking and writing about this central idea himself as his mind was so immersed in current and evolving questions and practical matters. Fortunately, more than most anyone else, Sir John worked so closely with Dr. Post from 2001 to 2007 on the question this book addresses. Thus, Sir John knew that he could invite Dr. Post to write a book on the central principles and concepts around Ultimate Reality and Unlimited Love that Sir John provided in his books, essays, and in conversational form with Dr. Post in order to allow us to engage in more probing reflection on the topic.

It is in this section of the book that we really see the authors' understanding of Sir John's sustained goals and his overall vision. From his earlier initiatives on "mind and ultimate reality," Sir John was prompted by the deeper quest for life's purposes based on spiritual perceptions that he hoped would be a blessing for others. Dr. Post correctly interpreted more and more clearly Sir John's high evaluation of the mind when he states that Sir John considered the rational human mind to be both a gift from God and a pinnacle of God's greatest love, as well as participatory in the mind of God and generative of "spirit". Sir John's appreciation for the mind shaped his continuously embraced devotion to the blessings of scientific rigor. This emphasis on rigor reflected his belief in the tremendous, almost unlimited potentials derived from the highest standards of science. Because the mind was so important to Sir John, he pursued with greater and greater anticipation the key centerpieces of "big questions" within the context of specific hypotheses and appropriate, clear and effective methodologies.

Thus, the book revolves around the format of asking perhaps the biggest question of them all, *Is Ultimate Reality Unlimited Love?* providing major and minor hypotheses, and pointing to plausible lines of empirical evidence that suggest future trajectories of rigorous investigation and even more questions. Pina Templeton noted that in the second section of this book, Dr. Post painstakingly looked for, and has collected, evidence from appropriate experts to support, or possibly challenge, what Sir John endorses in the form of truly incremental discoveries that serve only to produce more and more "big questions."

Finally, the third section is perhaps an easier, but nevertheless, an intriguing collection of statements from theological and scientific friends and from members of Sir John's family, to humbly seek and thus discover what they have learned from Sir John on the topic of love, unlimited love, and ultimate reality. The sources of such "learning" include not only recollections of diverse personal conversations with him but also insights from expanding insights from the reassessing of Sir John's printed or recorded writings.

Sir John clearly hoped that earlier "big questions" might creatively evoke new potentialities. For example, while this book is a compilation of Sir John's thought on ever expanding love—including greater substances framed as "ultimate reality"—it is also a "conversation" with Dr. Post whereby he and Sir John both *intentionally* meditate on wider and wider "exploratory concepts" and unanswered questions.

As a person of so many accentuated questions, Sir John most fully kept returning to Unlimited Love as the motivating force that energized his life. For Sir John, part of this love requires a full openness to truth in the form of *open-minded science*. Therefore, Dr. Post, in recognizing and continually expanding on Sir John's own words of "Ultimate Reality" has intentionally framed in this book spiritual, religious, and scientific perspectives as to why that "Ultimate Reality" may indeed be nothing short of "Unlimited Love" to be understood with greater clarity and rigor in the future. Dr. Post grasped a key element in Sir John's spiritual and scientific quest, and Sir John required him to grow and develop even to the point of writing this book several years after Sir John's passing.

Sir John had big dreams. He wanted to contribute positively to endurable, but also humble empowerment of individual persons and culture overall. He wanted to create new knowledge about God and the world of the spirit, and he wanted especially for such endeavors to foster gratitude and shifting priorities from "self" to "others". He envisioned that discovery would beget questions that would beget discovery that would beget more discoveries and so on for years to come, as we seek to understand ourselves, our culture and our universe."

Sir John believed that we know only a very small amount about the nature of God and the Spirit. He assumed that to make progress in learning – more people would have to be humble about their beliefs so that they would humbly but consistently be open to new discovery.

He believed that science had much to contribute to well-disciplined, structured questions as being essential to fundamental breakthroughs in "mind and spirit."

Finally, this book makes clear that Sir John saw Jesus as first a human reality and as a prototype of Unlimited Love. As Sir John himself grew older and also wiser through questions, he came to believe that this Unlimited Love might, with encouragement and prompting, also increasingly become a central and pivotal concept in other major world religions as a result of probing mental, moral and spiritual stimulations and continuous promotion. He asked if eventually science would converge with the great spiritual wisdom that underlying all of visible reality is a deeper Ultimate Reality of Unlimited Love. This discovery would truly give rise to spiritual progress on earth.

John M. Templeton Jr., MD
Josephine Joan Templeton, MD
February 2014

Preface

WRITING IN HOMAGE

L OVE WITHOUT LOYALTY, like love without laughter, makes no sense. I write in loyalty to Sir John Marks Templeton (1912–2008), who affirmed the centrality of deeply unselfish love in any well-lived life, and who was wonderfully bold and contrarian in asking if "Pure Unlimited Love" is really the energy matrix underlying the universe. I accepted an invitation from him to write this book on his behalf. The invitation came through Sir John's son, Dr. John M. Templeton, Jr, MD, who is everything that any father could hope for as he works diligently around the clock as president of the Templeton Foundation. Dr. Templeton was at his father's side in the days before his passing on July 8, 2008 in Layford Cay, Nassau. There Sir John mentioned to him that he wished he had the time left on earth to write one more book, one with the title *Is Ultimate Reality Unlimited Love?* They discussed the prospect of my writing it for Sir John consistent with his previous writings on the subject, with our conversations and correspondence over a number of years, and with our shared interest in the concept of Ultimate Reality ("God") as a matrix or substrate of Unlimited Love underlying and sustaining the visible universe.

Fortunately, Sir John left behind many statements on both Unlimited Love and on Ultimate Reality, making such a book possible. I thought about this project off and on for several years after Dr. Templeton brought it to my attention. I hestitated to write because it seemed presumptuous, but finally concluded that it is a good thing to honor the last request of a spiritually noble man.

As I developed this book, I made every effort to remain accurate with regard to Sir John's bold metaphysical ideas, drawing heavily on passages from his writings and from interviews with thought leaders and personal friends who understood him best. In preparing this material, I reread all of Sir John's books for the third time in the summer of 2011. The first time I read them was in the summer of 2001, when founding the Institute for Research on Unlimited Love, so named by Sir John, and the second was in the summer of 2008, after Sir John passed away and I was invited to become a trustee of the John Templeton Foundation. I could also draw on eighteen personal conversations with Sir John on the topic of Unlimited Love, which began in 1994, as well as on a number of fascinating letters from him challenging me to expand research into Ultimate Reality as Unlimited Love.

Sir John often emphasized that he was less interested in studying human love than the love that made humans. After one meeting he sent me a copy of his book titled *Pure Unlimited Love* with his business card attached to the front cover indicating a date of "31-8-01" in red pen with the note, "Mail to Stephen G. Post—Read pages 32 to 36 please." Those pages constitute a brief book chapter titled "Can Love Be an Eternal Universal Force More Potent Than Gravity, Light, or Electromagnetism?" This was just after I had founded the Institute with his support (www.unlimitedloveinstitute.com). The implication of his note was loud and clear: Be bold and ask the biggest questions possible about Ultimate Reality itself.

This is Sir John's "last book," and I am his helper. The title *Is Ultimate Reality Unlimited Love?* may cause some to do a double take. But Sir John liked such bold titles, and he gravitated toward them as if to make statements to the world that no one could mistake—clear as a bell and totally true to his convictions and interests, come what may.

As surrogate writer, I must state that I myself *do* find the idea of Unlimited Love as Ultimate Reality scientifically plausible and spiritually necessary. Otherwise I could not write this book honorably or well. Moreover, I hope with Sir John that one day in a future decade, as scientific methods and models accelerate knowledge exponentially, this perennial mystical idea will be demonstrated and proven true. On that day, in astonishment, the world will become truly joyful. This is ultimately the very essence of what Sir John meant by "spiritual progress."

When I met Sir John for the fourth time, in April 1996, he was sitting in the lobby of the Lansdowne Resort in Leesburg, Virginia, where he was attending a conference on Spiritual Interventions in Clinical Practice facilitated by our mutual friend Dr. David B. Larson, MD, president of the National Institute for Healthcare Research, for which I served as a member of the board of directors. Dave suggested that I go over and speak with Sir John about *agape* love. I walked over, and Sir John warmly asked me to sit down in the chair next to his. "I am curious about your life and interests," he said. He was so unassuming, so interested and concerned, and so undistracted in his listening and conversation. The next hour passed by quickly as we found common interest in the connection between unconditional and universal *agape*, and healing and recovery from illness—especially mental illness. He smiled through the hour, and he asked me as many questions as I asked him. This was a man from whom I wanted to learn. Yes, he really was so incredibly unassuming, like a good neighbor would be. He disciplined his mind to make room for the practice of love in every encounter.

For Sir John, the spiritual idea that Ultimate Reality is Unlimited Love constituted the most profound insight ever discovered by conscious humanity, and the most liberating by far. He understood this as revealed most fully in Christ, but he greatly respected its presentation wherever it might be found across the world religions. He liked the language of "Ultimate Reality" because the word "God" has been undermined by silly or destructive human anthropomorphic projections, ranging from a white-bearded old man to a sociopath trying to annihilate entire peoples. He liked the language of "Unlimited Love" because the Greek word *agape*, which amounts to the same idea of an unconditional and completely universal love for all people without exception, is so closely associated with Christianity that people from other spiritual traditions might feel excluded by it. Sir John had faith in Unlimited Love as Ultimate Reality, and he hoped for a day when this faith would be substantiated by what science learns about the underpinnings of our universe. That is the day when, as Paul wrote, we would see all truth, and know it as one.

Certainly Sir John himself intuited that behind the appearances of reality there lies a deeper matrix of Ultimate Reality that is constituted by the energy of Unlimited Love, but this never meant to him that scientists who seek support from the John Templeton Foundation needed to accept this theo-philosophy of the universe. And yet he clearly hoped to find pioneering researchers willing to ask such an audacious question as *Is Ultimate Reality Unlimited Love?* He often said that "maybe in a hundred years" science will have the methods and tools needed to truly investigate this biggest question of them all. He envisioned a day when great science and the perennial metaphysical tradition of Ultimate Reality found in all the world's great spiritual traditions might achieve a convergence around this matrix of Unlimited Love, and thus energize spiritual progress. This is a bold vision that few people can easily conceptualize this materialistic era, and thus the great challenge for the Templeton Foundation over the years will be to stay true to Sir John's intent.

Let us all take note of this passage from Sir John, which resonates with many others: "*Agape* love means feeling and expressing pure, unlimited love for evey human being without exception. Developing such a divine ability has been a goal for me for almost all of my eighty-six years on earth" (1999, 1). Clearly the life of love was the purpose of life, and as we shall see, like all the great spiritual thinkers, he hypothesized that the force and energy of Unlimited Love underlies the universe.

PART 1

*Why This Idea Meant So Much
to John Templeton*

An Analysis Based on His Writings and
Intellectual History Since Youth

Chapter 1

Sir John's Biggest Question

AN INTRODUCTION

I AM WRITING A BOOK with a title of Sir John's choosing as he had hoped to write it for himself before his health finally failed at the age of ninety-six. My hope in part 1 is to unveil his core theme of *Unlimited Love as Ultimate Reality* more or less as Sir John might have done so. I focus on his published writings, and draw on his letters to me on the subject over a seven-year period (2001–2007), as well as closely related conversations. In part 2, I focus on three intersecting axes of plausible scientific evidence for the hypotheses of *Unlimited Love as Ultimate Reality*, each of which I discussed with Sir John because I found them to be so central to his writings on the topic and queried him about them. In part 3, more than twenty scholars and associates or family members who knew Sir John generously reflect on what the concept of Unlimited Love as Ultimate Reality meant to him or on how he tried to live a life of love.

Throughout the book every effort has been made not to diminish or retreat from the full contrarian boldness of Sir John's mind, for he considered mere human love to be frail and unreliable, preferring to write often of Unlimited Love as the very Ground of Being (a.k.a. "God") that underlies all of reality. He meant this substantively, not symbolically or metaphorically. Sir John often asked me to investigate a love that is infinitely greater than our human love. He never intended to write a book on mere human love, for while he appreciated this asset, he understood such love to be limited with regard to duration, range, consistency, purity, and wisdom. Therefore, he saw "spiritual and religious progress" in our coming to a greater awareness of, and connection to, not human love but the love that made humans—that is, with the Ultimate Reality that is Unlimited Love and the matrix of being.

Sir John wrote me these words in a letter dated August 3, 2001, words that I know he thought deeply about and felt to be crucial for spiritual progress:

> I am pleased indeed, by your extensive plans for research on human love. I will be especially pleased if you find ways to devote a major part, perhaps as much as one third of the grant from the Templeton Foundation, toward research evidences for love over a million times larger than human love. To clarify why I expect vast benefits for research in love, which does not originate entirely with humans, I will airmail to you in the next few days some quotations from articles I have written on the subject.
>
> Is it pitifully self-centered to assume, if unconsciously, that all love originates with humans who are one temporary species on a single planet? Are humans created by love rather than humans creating love? Are humans yet able to perceive only a small fraction of unlimited love, and thereby serve as agents for the growth of unlimited love? As you have quoted in your memorandum, it is stated in John 1 that "God is love and he who dwells in love dwells in God and God in him."
>
> For example, humans produce a very mysterious force called gravity but the amount produced by humans is infinitesimal compared to gravity from all sources. Can evidences be found that the force of love is vastly larger than humanity? Can methods or instruments be invented to help humans perceive larger love, somewhat as invention of new forms of telescopes helps human perceptions of the cosmos? What caused atoms to form molecules? What caused molecules to form cells temporality? Could love be older than the Big

Bang? After the Big Bang, was gravity the only force to produce galaxies and the complexity of life on planets?

Sir John wanted to devote at least one third of his grant to support
investigations into a love "over a million times larger than human love."
Anything less would be an act of human arrogance. Two months later,
on September 1, 2001, he repeated by letter, "Unlimited love may be
billions of times more vast than any one temporary species on a single
planet can yet comprehend."

Dear readers, Sir John was not at all interested in writing a book
titled *Human Love as Ultimate Reality*. He was one who knew, both by
introspection and by observing human behavior, that human love is a
very limited enterprise, and human nature a very mixed bag indeed.
Although he condoned investigating it, he also counseled that our investigations focus to a very significant degree not on human love at all but
on Unlimited Love. I found this exciting because I agreed with his analysis, although in the modern secular university situated in a materialistic
era the very idea of studying Unlimited Love would prove challenging.
Sir John directly chided me in a nice way for supporting one small study
on non-human primates, stating, "This is not just very interesting to
me. Let's focus on Unlimited Love."

Sir John was always most curious about three areas of evidence
related to Unlimited Love as Ultimate Reality. First, he wanted to know
quantitatively how many people experience this Unlimited Love that
seems to invade and permeate their awareness, and how this experience
affects their behavior. He wanted hard numbers, although he was also
keenly interested in the qualitative understanding of this experience. A
decade later, after I finally assembled the right research team with the
immense help of the distinguished sociologists Matthew T. Lee and
Margaret M. Poloma, we conducted a national survey showing that an
estimated 80 percent of American adults self-report an experience of
God's love (see www.theheartofreligion.net) that enlivens their benevolence and is perceived as emotionally healing. Second, Sir John wanted
to know if people who love their neighbor have benefits with regard to

happiness and health, and if these benefits are amplified when such love is prompted by a sense in the giver of having a participation in Unlimited Love and serving as a conduit for it. Third, Sir John wanted to know if science can objectively demonstrate through physics and cosmology that the idea of Unlimited Love as Ultimate Reality is at least plausible with respect to its lying at the very origin and continuing essence of all that exists in the universe as the underlying matrix of being.

In other words, Sir John did not wish to leave the evidence with human experience alone, but he wished to extend it to the objective essence of the universe in the form of reality itself. (These three areas of research constitute part 2 of this book.) This was fine with me, although it flies in the face of a materialistic sensate culture with ruling scientific paradigms locking out such metaphysical possibilities. Like Sir John, I am a realist about human nature and doubted that we as a species could ever much amplify our love for all humanity without a dramatically increased awareness of the Unlimited Love that made and sustains all that is. Sir John's own lines about the presence of God in the universe will hopefully demonstrate to the reader just how intent he was on the third area of evidence. These lines are taken from his *Riches for the Mind and Spirit*:

> When man becomes humble in his approach to God, then he can think and speak in this way:
>
> Billions of stars in the Milky Way are upheld in the dynamic embrace of God's being, and He is much more. Billions and billions of stars in other galaxies are creatively sustained in God in the same way, and He is much more. Time and space and energy are all included within the power of God's presence, and He is much more. Men who dwell in three dimensions can apprehend only a very little of God's multitude of dimensions. God infinitely surpasses all the things seen and also the vastly greater abundance of things unseen by man.
>
> God is the only ultimate reality—all else is fleeting and contingent. The awesome mysteries of magnetism, gratitude, joy,

and love are all from God himself, and He is much more. Five billion people live on earth and live and move and have their being in God, and He is much more.

Untold billions of beings on planets of millions of other galaxies are what they are in God, and He is much more.

God is beginning to create His universe and allows each of us His children to participate in small ways in this creative evolution.

God is infinitely great and also infinitely small. He is present in each of our inmost thoughts, each of our trillions of body cells, and each of the wave patterns in each cell. God embraces all of us within the presence and power of His being, but we are a very little of all that subsists in Him. (1990/2006, 204)

Sir John would ask if love could be older than the big bang, and if we could seriously study a love "which does *not* originate entirely with humans." Sir John wanted to think big, think cosmic, think Ultimate Reality. And because I share his larger spiritual intuitions, I was going to write this book for him.

In 2001, my doctoral dissertation director, James M. Gustafson of the University of Chicago, under whom I had long before (1982–83) written a dissertation reconciling *agape* (God's) love with the flourishing of the giver, kindly advised me by phone not to undertake this adventure because it "was beyond the acceptable paradigms and might well destroy your career." "You and Sir John may be right," he said over from New Mexico, "but the academic world will not tolerate such paradigms for a century or more, if then." But by December 2013, Professor Gustafson congratulated me by phone on the Institute and on this book. The idea of *Unlimited Love as Ultimate Reality* is actually not at all a strange idea for anyone versed in the history of world religions. For example, here is a quote from Hazrat Inayat Khan (1882–1927), who was born in India and ventured to the West in 1910 to found the international Sufi movement in 1918. He wrote,

And the love of God is that which is the purpose of the whole creation: if that were not the purpose, the creation would not have taken place. As the whole creation is from God, then it is of God. If it is of God, then it is a manifestation of love, and the manifestation of God is purposed to realize perfection in love. (in Rose 2007, 215)

Sir John greatly appreciated Sufi metaphysics, and befriended the great contempory Islamic scholar of divine love, Seyyed Hossein Nasr, University Professor of Islamic Studies at George Washington University and a graduate of the Massacusetts Institute of Techology. Islam and Hinduism have more easily sustained this metaphysical tradition of Ultimate Reality as Unlimited Love than other great faiths, but it is essential to each and every one of them with regard to their great mystical exemplars and deeper spiritual thought.

No matter what we do, "God loves us all equally and unceasingly," wrote Sir John (2000a, 127). It is said that God can do anything. Sir John would have said that this is not true. There is one thing that God cannot do. God cannot reject anyone, no matter what difficult and even despicable a life they may have lived, because God's Unlimited Love is too radically big for that. People feel sometimes that God cannot love them because of a mistake they made, a poor decision, or some bad habit or personal history. They feel unlovable and even hated by God. But the one thing God can never do is not love anyone. God's Unlimited Love is millions of times deeper and more unchanging than anything we human creatures have within us by nature alone.

This radically unconditional, universal, and unchanging love made Sir John thankful, and this was at the center of his interest in gratitude, which always went beyond mere interpersonal gratitude to an appreciation for a God of Unlimited Love. His son Dr. Templeton, along with other relatives, reports that at about the age of ten John Templeton was already speaking of how much gratitude he had for a God who could love him, and love us all, overlooking our many faults. Young John Templeton preferred Thanksgiving to any other holiday for this reason, and later in life, during the 1960s, began to write Thanksgiving cards to all his family and friends.

Sir John thought that humanity has not yet begun to explore the infinity of potential in this Unlimited Love. Such love can free us at last from the shadowy myths of fear-based gods and it can open the doors to a future free of the conflict and violence that arises from our tendencies toward retaliation, domination, exclusion, vengeance and most of all from the arrogance that underlies these evils. Hence, he created Humility Theology, focusing on how little we know about God. Unlimited Love for Sir John represents the greatest leap forward in consciousness and progress imaginable.

In speaking of Unlimited Love as Ultimate Reality, Sir John was also referring to the most significant grounding for human freedom conceivable. A God who is love is not the God of slavish, fearful obedience. In the place of coercion is the voluntary loving relationship of freedom. Love is the very basis of freedom, for influence is exercised not through superior force but instead through truth telling, persuasion, and attraction. So it is with God's love. Knowing that human nature tends to choose security over freedom, Sir John considered freedom an aspect of God's love that is bestowed upon us and that we should accept. Freedom is not rightly understood by analogy to a dog wanting to chew through its leash. In fact, the dog will stop chewing whenever a large plate of food is placed before it. No, for Sir John freedom is an inherent element in the eternal part of the human mind that is creative and loving, and participatory in Infinite Mind. In other words, freedom has its origins in God.

A ROUTE 80 MOMENT

As alluded to in the Preface of this book, I was pleased to undertake this responsibility to write on Sir John's behalf, but did so only after considerable prayer in July 2011. With a bit of bright sunrise inspiration at the Delaware Water Gap on Route 80 driving west to attend a funeral in Cleveland, I pulled over and called Dr. Templeton on my cell phone to tell him that, indeed, I would like to give this book every effort. I pulled over because a retired pediatric trauma surgeon like "Dr. T" would not like to see me on my cell while driving. My qualifications for this project are (a) an absorption in Sir John's writings over a span of two decades, (b) personal interaction with Sir John on the theme of Unlimited Love for more than a decade and as founder and president of the Institute for Research on Unlimited Love that he supported, (c) interactions with his family and friends as associated with the John Templeton Foundation, and (d) a personal interest in the topic of God's love that led to the writing of my Sixth Form thesis on the topic at St. Paul's School in New Hampshire in 1969, and that has defined a career spanning forty years that has not deviated an iota from this interest.

Dr. Templeton was clear in not only affirming my commitment but also in stating that this was one of the best things I could do as a trustee of the Templeton Foundation to encourage future generations of individuals associated with it to take his father's thought world with the seriousness that it deserves, even if it is difficult for them to conceptualize Divine Mind, Ultimate Reality, and Unlimited Love. Dr. Templeton wanted to avoid any drift away from Sir John's full intent in founding the Foundation. So we firmly agreed on the project in that ten-minute conversation. I was careful to put my cell phone down before heading west again on that great highway.

This book is not a formal intellectual biography, although an assessment of major influences on Sir John's thought is offered. Rather, the book is an attempt to write more or less what Sir John would have wanted to write himself about his core idea had he lived a little longer. I am not writing as a historian or a biographer, because Sir John would not have written the book this way. Rather, I am working with Sir John's major publications and putting myself in his shoes as best I can. Hopefully, this book makes the case he wanted for his core assertion in a manner that appeals not just to professional theologians, philosophers, and scientists but to all those who knew him and loved him, and to the informed lay public that deserves to understand the deep spiritual and intellectual authenticity of the man who was both one of the twentieth century's greatest investors and the founder of the Templeton Prize for spiritual progress.

In the final analysis, Sir John wanted to transform our increasingly decrepit and destructive culture by reanimating the great perennial spiritual and metaphysical truths, Unlimited Love as Ultimate Reality being chief among these. Thus, I write as his fiduciary, and stress his metaphysical commitments and biggest questions, none of which can be contained within or appreciated by mere evolutionary naturalism, for Sir John was definitely a nuanced substance dualist who believed that Mind is not merely an epiphenomenon of matter but also has a primacy of it own. In a nutshell, Sir John rejected "ontological reductivism," that idea that Mind as an order of being can be explained as "nothing but" matter in the form of brain chemistry and neurons.

Can we please take this ontological richness seriously? The readers of this book might want to read a work by Thomas Nagel, who many consider to be among the several greatest philosophers of mind in the world. Nagel, who spent most of his career at Princeton, has written a book entitled *Mind & Cosmos: Why the Materialist Neo-Darwinian Conception of Nature is Almost Certainly False* (2012). He rejects the current materialist "orthodoxy" and asserts that our "successors will make discoveries and develop forms of understanding of which we have not dreamt" (2012, 3). Nagel writes that "our secular culture has been browbeaten into regarding the reductive research program as sacrosanct, on the ground that anything else would not be science" (2012, 7). He simply does not think that consciousness and mind emerge from matter, or that the universe could emerge without an underlying Mind. Nagel is not a believer in "God" or "theism," or "intelligent design," although he does think that the philosophers of intelligent design make some important points and are unfairly excoriated. He is rather a classical metaphysical idealist in the tradition of Plato: "The view that rational intelligibility is at the root of the natural order makes me, in a broad sense, an idealist—not a subjective idealist, since it doesn't amount to the claim that all reality is ultimately appearance—but an objective idealist in the tradition of Plato and perhaps also of certain post-Kantians, such as Schelling and Hegel, who are usually called absolute idealists" (2012, 17). He asserts that "the intelligibility of the world is no accident" (2012, 17), and that "conscious mind" is prime and original in the universe, rather than secondary and evolved. I make no claim here that Nagel would associate this original prime Mind with some form of classical theism, or with Unlimited Love, but he has clearly joined the ranks of the metaphysical idealists who do not think that random chance and materialism can explain the universe or our human minds. I will assert that if a philosopher as preeminent as this thinks that logic brings us to an original Mind at the source of things, then Sir John deserves to be taken with great seriousness as he was ahead of his time. Increasingly, philosophers of mind are concluding that nothing in the material universe can be the source of our minds, however much our minds may be interwoven with our brains.

Chapter 2

The Spiritual and Intellectual Roots
of Sir John's Biggest Question

AT THE AGE of eighty-six Sir John Templton looked back on his life and wrote these words: *"Agape* love means feeling and expressing pure, unlimited love for every human being with no exception. Developing such divine ability has been a goal for me almost all of my eighty-six years on earth. (1999, 1). Where did this purpose come from? How did it come to shape Sir John's most central metaphysical question, *Is Ultimate Reality Unlimited Love?*

To understand Sir John's ideas on this question it is necessary to clarify the two formative spiritual-theological influences that shaped Sir John's thought from youth. Sir John was a Presbyterian with a very realistic view of human nature, and yet he was also inspired by the Unity School of Christianity, which with its New England transcendentalist roots swept into Missouri and made its way to a young Sir John in Franklin County, Tennessee, via radio broadcasts and a little monthly reader called *The Daily Word.* In many ways, Sir John's lifelong commitments to spiritual progress emerged at the synergistic interface of these two influences. He had the realism of Jonathan Edwards and the hope of a Ralph Waldo Emerson.

THE CUMBERLAND PRESBYTERIAN CHURCH

Imagine a man of eighty-six honestly and humbly stating that his purpose of living for "almost all of my eighty-six years on earth" was to abide in the ways and power of Unlimited Love. Sir John indeed did feel this purpose or calling very early on. As a teenager in 1927, for example, at the age of fifteen, he became the superintendent of the Cumberland Church Sunday school in Winchester, Tennessee. His beloved mother, Vella, an elected elder of the church, encouraged young John to take on this administrative role. The young Templeton also taught Sunday school lessons, and took a special interest in hymns and biblical passages having to do with the theme of *agape* love.

Dr. Templeton reports the story of how an old Tennessee farmer asked him once, "Are you John Marks's boy?" He added, "That John Marks, he was born old!" meaning that he had a wisdom and purpose about him early in life. So it is that Sir John's retrospective self-assessment of having a loving purpose in life all of his years seems to have considerable local anecdotal support.

On June 24, 2012, I had the opportunity to attend a Sunday commemorative service at the Cumberland Presbyterian Church in honor of John Templeton's one hundredth birthday four years after his passing. The service comprised hymns, prayers, and scriptural passages that Sir John had personally identified to the Rev. Dr. Jonathan Clark as especially meaningful to him. Many pertained to love and gratitude. That bright Tennessee morning I heard Sir John's granddaughter, Heather Templeton Dill, read "Sir John Templeton's Personal Prayer of Thanksgiving" to those assembled, perhaps half of whom were family members or those associated with the Templeton Foundation. The prayer, written in the 1960s, is a meditation on gratitude for the love of God. It is an important marker of Sir John's core spirituality and thus must be included here in full:

> Almighty God, our loving Heavenly Father, through faith and the Holy Ghost, we are totally one in unity with thee. Thou art always guiding us and inspiring us to the right decisions in family matters, in business matters, in health matters, and especially in spiritual matters. Dear God, we are deeply, deeply grateful for thy millions of blessings and millions of miracles that surround us each day. We are especially grateful

for thy healing presence, which gives us long and useful lives in which to love thee more and more and to serve thee better and better. Dear God, help us to open our minds and hearts more fully to receive thy unlimited love and wisdom and to radiate these to thy other children on earth, especially today and all this year. Dear God, we thank thee for blessing and healing each of our families and friends and for helping each of us to be better and better Christians. We thank thee for thy miraculous and continued blessing, guidance, and inspiration of our careers and daily work to serve others in business and churches and charities, so that all of these will be more and more in accord with thy wishes, O Lord, not ours. We listen and obey and are grateful. We thank thee for our redemption and salvation and for the gift of thy Holy Ghost, by grace, which fills us to overflowing and increasingly dominates our every thought and word and deed. To thee we pray, in the name of thy beloved Son, whom we adore and seek to imitate, our Saviour and our God, Christ Jesus. Amen. (1990/2006, 24–25)

This prayer was sent out in the form of a family Thanksgiving card. The reader is struck by the spiritual depth in this prayer, as well as by its eloquence and sincerity. It could only have been written by an individual who took God seriously and for whom Unlimited Love is the ultimate interest. This is not a prayer about human love. Rather, it is a prayer about a "Dear God" who can "help us to open our minds and hearts more fully to receive thy unlimited love and wisdom and to radiate these to thy other children on earth, especially today and all this year." Indeed, it is the first time that Sir John explicitly used the term "Unlimited Love" so far as I am aware.

In 1950 Sir John accepted a position on the board of trustees of Princeton Theological Seminary as a ministry to his church where he had many Presbyterian friends. There he befriended Rev. Dr. Bryant Kirkland, then minister of the Fifth Avenue Presbyterian Church in Manhattan, and Dr. James L. McCord, then president of the seminary. I had the opportunity to get to know Dr. Kirkland fairly well before he passed away on May 5, 2000, still Sir John's loyal friend and close advisor. John was also a close friend of Dr. Norman Vincent Peale, and he enjoyed friendships with Dr. Robert Schuller, another Christian positive thinker. Rev. Schuller, who I knew fairly well from appearances on his *The Hour of Power*, often spoke of how close he felt to Sir John. David Myers of Hope College, Dr. George Gallup Jr., and many leading lights in the Reformed Protestant tradition remained close to Sir John over the course of his life. On the whole, he tended toward those who appreciated the power of positive thinking and affirmation as one way of keeping a wayward human nature focused on higher spiritual principles. Sir John's Presbyterian roots were always a part of his life, and he honored them to the end, even as he became more and more eager to learn about all the religions of the world as they also had articulated insights into Unlimited Love.

On June 2, 2012, only several days before this commemoration service at the Cumberland Presbyterian Church, I had spent some time in Sir John's office in the Templeton Building in Lyford Cay in the Bahamas, where he lived and worked for the last forty years of his life. I looked up and saw on the wall behind his brown wooden desk a large framed page with a calligraphy-style set of passages from the New Testament, Luke 6:27–38 (the Sermon on the Plain). For those unfamiliar with this passage, it too is worth quoting here because time and again Sir John referred to it and took guidance and inspiration from its words daily:

> But I say to you who hear: Love your enemies, do good to those who hate you, bless those who curse you, and pray for those who spitefully use you. To him who strikes you on the one cheek, offer the other also. And from him who takes away your cloak, do not withhold your tunic either. Give to everyone who asks of you. And from him who takes away your goods do not ask them back. And just as you want men to do to you, you also do likewise.
>
> But if you love those who love you, what credit is that to you? For even sinners love those who love them. And if you

do good to those who do good to you, what credit is that to you? For even sinners do the same. And if you lend to those from whom you expect to receive back, what credit is that to you? For even sinners lend to sinners to receive as much back. But love your enemies, do good, and lend, hoping for nothing in return; and your reward will be great, and you will be sons of the Most High. For He is kind to the unfaithful and evil. Therefore, be merciful, just as your Father also is merciful. Judge not, and you shall not be judged. Condemn not, and you shall not be condemned. Forgive, and you will be forgiven. Give, and it will be given you; good measure, pressed down, shaken together, and running over will be put into your bosom. For with the same measure that you use, it will be measured back to you.

These words explain the Sir John who so often extolled us to give more freely and wisely to our neighbor in need, and not to be limited by any law of reciprocity. He saw expectations of payback as a complete antithesis to the free overflowing fountain of our serving as conduits of God's Unlimited Love.

Let us return again to his words quoted above, "almost all of my eighty-six years on earth." They suggest that *agape* love can be everyone's purpose from the earliest years with good nurturing in family and community, even before a child can comprehend such a noble purpose. Sir John owed so much to his parents, his community, and his church. One finds no mention in the above passage of his having had that single intense transformative spiritual experience of the newly repentant narcissist being overwhelmed by the life-changing light. Sir John greatly respected the dramatic experience of rebirth as one window to the divine, but his spiritual journey was less choppy. His profound awareness of Unlimited Love and of Divine Mind was very real but smoother than for some others. He experienced good Christian nurture in his faith community, and he stuck with the values and virtues that he learned early on. All this Cumberland experience was hugely formative, but so was the Unity school of Christianity, his second great spiritual influence.

THE UNITY SCHOOL OF CHRISTIANITY: AMERICAN TRANSCENDENTALISM VIA MISSOURI

At the Unity school of Christianity United Kingdom Conference in 1995, Sir John presented an address titled "Understanding: Acceleration in Spiritual Information." He began, "For more than seventy-five years, I have been reading the works of Unity school of Christianity, especially the writings of Charles and Myrtle and Lowell Fillmore. The Unity school viewpoint is especially attractive because of its willingness to be open and receptive and nonopposing to the writings of other denominations and religions." There is no doubt about his having absorbed Unity thought from youth. His mother, Vella also stimulated his spiritual interest; she received at the Winchester family home a small monthly booklet of spiritual affirmations called *The Daily Word*, published by Unity. This publication is still available and popular across the United States and abroad. It comprises brief daily affirmations, one per page, followed by quotes from world scriptures and a few words of inspiration. Sir John read each booklet with enthusiasm in his teenage years and developed a great admiration for Charles Fillmore and, through him, for the themes of New England transcendentalism.

As a Tennessee teenager during the 1930s, Sir John discovered a truth through Unity that arches across the ages in a trajectory from the philosophers of antiquity such as the Roman Marcus Aurelius, who wrote, "Your life becomes what you think," and carries forward to Charles Fillmore's statement, "Thoughts held in mind produce after their kind." Indeed, Sir John cites both these passages in *The Essential Laws of Life*. Could anyone seriously doubt that our thoughts have immense implications for how we focus our energies and creativity in every domain of life? This emphasis on the power of brief affirmations cultivated through daily meditation entered Sir John's young life through Unity's *The Daily Word*.

Briefly, Unity was cofounded by Charles (1854–1948) and Mary Caroline "Myrtle" Page (1845–1931) Fillmore in Kansas City, Missouri, informally in 1889 and formally in 1903. Charles was born in Minnesota, and Myrtle in Ohio. The informal beginnings of Unity go back to the 1880s when the couple, struggling with economic and health challenges, found solutions in American transcendentalist spiritual currents of the time such as the writings of Emanuel Swedenborg, Ralph Waldo Emerson, Theosophy, Hinduism, Quakerism, New England's "mind healer" Phineas Parkhurst Quimby (1802–1866), the Bible, and Divine Science. Fillmore believed that spiritually progressive people should welcome relationships with people of all religious persuasions and that they should accompany creeds with simple "spiritual principles." Fillmore adapted a phrase found in the Hindu Upanishads, a sacred text, that "What you think you become." If we think good we become good, and if we think bad we become bad. He considered thought as the source of creativity and as real as matter.

This particular aspect of Fillmore's "new thought" had a profound influence on Sir John, and underlies his "laws of life" focus on short affirmations that focus the mind in creative directions. For Sir John as for Fillmore, mastery of one's thoughts is mastery of one's soul and fate. By practicing concentration (*Dharana* in the Sanskrit) we achieve a unity with our thoughts that enable us to unleash the deeper levels of energy that lie within us, for the soul is one with God.

Historians of American religion view Unity as a Missouri outgrowth of the New England transcendentalist movement of the latter half of the nineteenth century. The Bostonian transcendentalists were an eclectic lot. Harvard Divinity School's Ralph Waldo Emerson, the thought leader of nineteenth-century transcendentalism, wrote of the Oversoul in terms of "Divine Mind" as the origin of all things, a concept very much the same as the Hindu Brahman and the Christian *logos* (i.e., God is *logos*, or a thought blueprint that manifests in creation). All things material are emanations of the One Mind and its energies. In the transcendentalist tradition, *mind* and *spirit* are used synonymously. Also, the human mind was deemed a very small piece of Divine Mind, and we have been given our minds so that we might love and participate in the continuing creativity of the Divine Mind that is in us; here the Hindu tradition exerted itself powerfully in Americanist spirituality. This omnipresent Divine Mind was deemed immanent, enfolding and interpenetrating all things. While prayer was emphasized, daily meditation was also stressed as a way of opening up the mind to awareness of the Divine Mind, within which the soul was included. Moreover, the self-control of the mind was deemed a sacred responsibility, made possible in large part by the use of brief spiritual affirmations repeated and inculcated into consciousness. Joy flows from the union between the human mind and the Divine Mind, which are of the same essence. All these ideas would become essential in Sir John's theo-philosophical worldview.

It is notable that Unity and New Thought figures that appear in Sir John's writings include the most influential Unity school thinker of her day, Dr. H. Emilie Cady (1848–1941), Emmet Fox, the Fillmores, Church of Religious Science founder Ernest Holmes, and other great New Thought minds. Let us take a brief illustrative quote fom Holmes's 1938 classic *The Science of Mind* (1998) to touch the essense of New Thought: "Man's mind is the Mind of God functioning at the level of man's understanding of his place in the Universe. Man contacts the Mind of God at the center of his own being. It is useless to seek elsewhere. 'The Highest God and the innermost God is One God.' Through the medium of Mind man unifies with the Universe . . ." (1998, 394).

In *The Varieties of Religious Experience* (1902/1982), William James pointed out the Hindu influence on the Americanist New Thought movement, in which Unity was a hub. James wrote of the emphasis on overcoming the barriers between the individual and the Absolute as the great mystic quest of the New Thought movement, which he considered a synergistic combination of the Gospels, Emersonian transcendentalism, Berkeleyan idealism, popular science evolutionism, "and finally, Hinduism has contributed a strain" (1902/1982, 94). But we must add Christian theology to the mix, the Bible, a bit of Quakerism with its Inner Light, and some focus on psychosomatic healing in the form of Mind Cure.

Fillmore imbibed all of these influences, and through his radio addresses from Missouri in the 1920s he influenced the young John Templeton in Tennessee, an avid listener, for life. In the 1960s Sir John would listen as well to the great Unity speaker Eric Butterworth, who he met at the Unity Center of Practical Christianity in New York City and whose Sunday services at Carnegie Hall were broadcast throughout the United States, in Great Britain, Europe, Africa, and the isles of the Caribbean.

Let me return to Sir John's office in Nassau and that visit on June 20, 2012. Symbolically, a moment of insight into the transcendentalist side of Sir John came to me in that quiet moment on Wednesday, as I observed the details. I spent time looking at the items on the wall, shelves, and surfaces, all of these unmoved since his passing in 2008. Sir John's much beloved daughter-in-law Pina was quietly working at his big brown wooden desk. By the well-worn leather couch where he took naps was a table, and on the table was a book by Quimby (1802–1866) with a page folded in over a passage about the healing power of God's love. Quimby was the New England idealist metaphysician, theophilosopher, and healer whose ideas were important to the Unity school and to all subsequent New Thought. Quimby's life and thought also gave inspiration to Mary Baker, founder of Christian Science, whom Fillmore regarded highly and whose writings clearly are in the mix of New England transcendentalism. Through Unity and its transcendentalist roots, Sir John was indebted, then, to the Hindu tradition that was so influential on Fillmore. Hinduism permeated all the synergistic spiritual circles of late nineteenth-century America, including the American transcendentalist movement in which Christian and Vedantic Hindu thought worlds synergized.

CO-CREATIVE PARTS OF A LARGER DIVINE MIND

For Sir John, the question of purpose is more or less equivalent to the question, why are we created? He wrote a book of this title with a minister from the Unity school of Christianity, Rebekah Alezander Dunlap.

The subtitle of their *Why Are We Created?* says it all: *Increasing Our Understanding of Humanity's Purpose on Earth* (2003). Here, in order to avoid any iota of ambiguity about the source of ideas, I concentrate entirely on the Introduction to this book, which Sir John penned independently and signed John Marks Templeton.

Sir John asserts the following pronouncement: "Twenty-five centuries ago, Xenophanes, and twelve centuries ago, Shankara, taught that possibly nothing exists independently of God, and that God is immeasurably greater than all of time and space, let alone the visible earth and its billions of inhabitants!" (2003, xi). His message is clear—we are not going to get very far along in our understanding of human purpose without taking God into account. So we are driven to ask what the purpose of God is. "Might a purpose of infinite intellect be to express itself in increasing varieties of lesser intellects?" he asked. Then he adds, *"Were human beings created to be agents of God's accelerating creativity? Can humans discover larger fractions of infinite intellect?"* (2003, xv, italics added). From conversations with him, and from his writings, it was clear that indeed he did think that each of us is created to expand in freedom the creativity and love in which we participate as drops in the ocean of Divine Mind. Sir John clearly thought that our minds, in their deepest eternal element, are God's love gifts, that they are interwoven at least in their most profound capacities in one larger Divine Mind, and that they confer upon us the purpose of extending novel creativity through our human agency as co-creators. This idea of the purpose of life is conceptually indebted to the great Indian philosopher Shankara, in whom Sir John had a considerable interest.

Shankara

Sir John made a good choice in picking out the great Shankara (788–820 CE). This Indian sage from Kalady lived only thirty-two years, but westerners consider him the greatest Hindu philosopher of them all. Why? He formulated the doctrine of *Advaita*, which refers to the identity of the Atman ("self" or "spirit" or "mind") and the Brahman (the "Supreme" or "Ultimate Reality, Infinite Mind" that is the origin and support of the phenomenal universe, the "Godhead" or "Absolute," the "Divine Ground of All Being"). In the Hindu tradition, the *Mahatmas* are "great souls" who have realized the union of the self and the Supreme. Shankara said that we are like individual waves on the ocean that cannot exist other than as part of the ocean itself, an idea elevated from the Upanishads and the Vedic canon.

The Vedantic school of Hindu philosophy is about Ultimate Reality (the term has its origins in Vedantic thought) and our relationship to it. This school is the largest and most dominant perspective in all of Hindu thought. A key group of scriptures known as the Upanishads—written in the form of a dialogue between a student and a teacher by many different authors starting in the seventh century BCE—constitute the final section of the Vedas. We know that Fillmore read these dialogues in depth, and was devoted equally to them and to the Bible. Two fundamental concepts are established in the Upanishads, the Atman and Brahman. The Brahman (not to be confused with the word "Brahmin," meaning priest) is consistently referred to as the primary ground of all being and existence. Sometimes Hindu philosophy views Brahman as impersonal and sometimes as personal and loving. Shankara, whom Sir John was interested in most, takes a personal view of Brahman as something like "Love-Mind."

Here in *Why Are We Created?* Sir John, then ninety-one years of age, was reflecting deeply on the ultimate nature of human purpose, and he was drawing on Shankara. He asks, to restate the question (italis mine), *"Might a purpose of the infinite intellect be to express itself in increasing varieties of lesser intellects?"* (2003, xv). He is stating here that the Infinite Mind of God flows through us all, dwells within us, and actually is constitutive of an element of our individual minds. Moreover, we are the creatures through whom Divine Mind expands its creativity. The Ultimate Reality wants to extend its creativity in novel and loving ways, and we are the agents of the co-creative extension.

In Shankara's philosophy, we find best articulated the perennial formula that Atman = Brahman. None other than Marcia Eliade at Chicago, and later visiting professor Joseph Campbell, explained this concept to me. Both used the analogy of sand. Each of our individual souls (Atman) is like a grain of sand. The whole mound of sand together makes up the Brahman, which shares bits of itself with each human being, but remains infinitely large. So the deep eternal essence of the human mind is of the same substance (in minuscule) as is the Infinite Mind. Thus, God exists in each one of us. The Hindu greeting *Namaste* is based on this concept and is translated, "The divine in me honors the divine in you." The idea of Atman = Brahman is the basis of the nonviolent tradition in Hinduism and all Hindu ethics. Of course, the Atman can be covered over by bitterness, anger, hostility, rage, jealousy, fear, resentment, hatred, and the like. But it is nevertheless the original nature and the "true self."

In Hindu thought, when we gaze outward at the starry heavens and the universe, we see a small part of the expanding emanation of the Brahman (consistent with big bang theory), and when we look inward at our Atman we meet the same Brahman within us. No distinction is made between our souls and Ultimate Divine Reality. Brahman, sometimes described as a universal force of pure love and pure thought, is all and created all. There is a piece of Brahman in everyone, and the Atman is not affected by emotions or the body's physical needs. Through the Atman that is a piece of Brahman, all of us are interconnected. The highest, Brahman, is no different in essence than the smallest, Atman. The power, love, intelligence, and creativity of the Brahman can be unbound in the Atman, so we are each capable of astonishing love and creativity that go far beyond the biological substrate of the mind. As Shankara taught, that which is everything is that which is your essence. All that exists is grounded in the Great Thought from which all that is emanates continuously.

In Sir John's very important but a tad neglected work *Agape Love: A Tradition Found in Eight World Religions* (1999), his chapter on Hinduism is in my view the most brilliant. He begins,

> Hinduism speaks of the self, or soul (Atman). It also speaks of Brahman as being the ultimate principle of the universe. The fundamental religious conviction that Brahman is Atman, or

that the self is ultimately inseparable from the whole, lays a firm foundation for agape in the Hindu context.

Because all human beings are in some sense one, and indeed because all of creation is one, the only way to treat others is with respect, kindness, justice and compassion. (1999, 45-47)

Aldous Huxley refers to Shankara's thesis as follows: "The Atman, or immanent eternal Self, is one with the Brahman, the Absolute Principle of all existence; and the last end of every human being is to discover the fact for himself, to find out who he really is" (1944/2009, 4). Huxley quotes Shankara for several pages—for example, "The wise man is the one who understands that the essence of the Brahman and of Atman is Pure Consciousness, and who realized their absolute identity" (1944/2009, 6). This is little different, writes Huxley, from the words of Meister Eckhart, "To gauge the soul we must gauge it with God, for the Ground of God and the Ground of the Soul are one and the same" (p. 14). All the mystics from all traditions are more or less describing the same ineffable experience of astonishing awareness that some part of the individual mind is completely one with God. The Quakers speak of the Inner Light, and that language can be very useful. In the eternal soul, this Atman, human freedom, love, creativity, and ingenuity have their source in God.

So what is the highest human purpose according to Sir John? The purpose of any human life is to be aware of our oneness in mind with the Divine Mind, and to serve as an extension of God's creativity and Unlimited Love. Sir John meant it when he wrote, "Were human beings created to be agents of God's accelerating creativity? Can humans discover larger fractions of infinite intellect?" (2003, xv). This is an idea with roots in the Hindu influences on Fillmore and Unity.

A GREEK SAGE

In his Introduction to *Why Are We Created?* Sir John mentions Xenophanes of Colophon (c. 570–475 BCE) in addition to Shankara.

Xenophanes was one of the most important pre-Socratic Greek philosopher-poets, most famous for satirizing the anthropomorphic images of the Greek gods. Xenophanes was critical of the religious views of his day because they were merely human projections—a position Ludwig Feuerbach took up in relation to modern Christianity. Xenophanes made an innovative critique of Homer and Hesiod, who attributed to the gods every conceivable human image and vice. The petty hatreds, jealousies, and hostilities of the gods, and all the nefarious actions that followed, such as theft, deception, adultery, and violence, would typically be considered matters for reproach in human behavior. In this sense, Xenophanes was very influential for Plato, who in the *Republic* urged that these gods be abolished because they only contribute negatively to human behavior. Xenophanes is generally thought to be among the first ancient philosophers of monotheism. He argued for a God beyond human form, who is eternal not born, and who is conceivably the whole of the universe.

Sir John appreciated anyone who based his philosophy of religion on the idea that whatever human beings may be thinking about God or the gods is "too small," and in this regard he identified with Xenophanes. I would also assert that Sir John entirely dismissed sociopathic anthropomorphic images of the God of Unlimited Love, and he clearly believed that because of these projections religions undermine the true God as they bring out not just the best but sometimes the worst in people.

In the famous Fragment B14–16, Xenophanes comments as follows:

> But mortals suppose that gods are born,
> wear their own clothes and have a voice and body. (B14)

And:

> Ethiopians say that their gods are snub-nosed and black;
> Tracians that theirs are blue-eyed and red-haired. (B16)

But perhaps most famously, he wrote,

One god greatest among gods and men,
Not at all like mortals in body or in thought. (B23)

And this one greatest god "shakes all things by the thought of his mind [*nous*]." This idea influenced Aristotle and many others, who followed Xenophanes in his idea of a divine intelligence underlying all of reality. Later Greek writers indicate that Xenophanes identified this one god with the entire physical universe. Sir John would ask, is God the only reality?

Xenophanes also influenced Anaxagoras (born c. 500–480 BCE), the first pre-Socratic philosopher to live in Athens and the father of Athenian philosophy. Anaxagoras claimed that *nous* (intellect or mind) was the cause of the entire cosmos in its movements, principles, and being. He also is widely associated in Greek thought with the original idea that the mind of God is in each of us, and he equated mind with eternal soul.

SIR JOHN'S BIGGEST QUESTION AS A HYPOTHESIS

Sir John asked the very biggest questions about love, ones that engage cosmology and physics, questions that are so big that they go right to the underlying purpose behind the universe and are rather astonishing to read: "Is love more universal than the universe? Could it be true that in a deep way, a purpose toward love really may govern the sun and the stars? The humble approach is about many questions. How little we know—how eager to learn! That is the humble approach" (2000a, 12).

Physicists with a more mystical bent debate these same truly big questions. Sir John greatly appreciated the cosmologists and mathematicians who speculated with ease on the ultimate nature of reality as infinite mind, consciousness, and energy. He felt that scientific progress eventually will find an answer to these deeper big questions, although it might take a very long time.

I have described Sir John as a spiritually nurtured Tennessee Presbyterian with a significant appreciation for the Unity school of Christianity, and as his years advanced he became more and more engaged with Unity and other New Thought ideas about Divine Mind. He dared to ask if Unlimited Love is the enduring and perfectly reliable matrix that constantly creates and sustains everything that exists, including the laws of physics. He dared to ask if our very minds might be parts of an infinite Divine Mind given to us like a drop of water in the sea as a "Love-Gift" so that we too might participate in co-creativity and love. He dared to ask if Unlimited Love could be the essence of an original, eternal, prime creative thought energy underlying everything that is energized and materialized across the universe. Far from assuming that mind emerges from matter and cells, he felt it somewhat more plausible that at the beginning of all things and underlying all things there is a prime or original Divine Mind from which the eternal aspect of each individual minds is given, such that each of our minds is in essence a location within a larger field of Divine Mind. Further, he considered it plausible that all energy and matter, and all the laws of the universe, derive from Divine Mind.

In keeping with his vision, Sir John Templeton quoted a particular mystical passage time and time again. It is from Paul's speech to the Athenians, where Paul asserted a much bigger God than any that they worshipped, a God who made the world, who is immanent in it, and who does not need to be searched for because this God is "not ever far from any of us." As Paul famously continues, "In him we live, and move, and have our being" (Acts 17:18).

Sir John had a favorite phrase, "Every person's concept of God is too small." He felt that too many people become absolutistic and arrogant about their little notions of God and therefore get stuck in the past; worse yet, these people are utterly unappreciative of others who have another idea of God that might be quite interesting. Moreover, small ideas of God have been developed over time without the benefits of modern scientific progress in areas such as quantum physics and cosmology, so they do not appeal to the modern enlightened mind. Sir John prescribed humility, eagerness to learn, appreciation of various spiritual traditions, and a God who is everywhere and in everything as well as plausible from the perspective of the physical sciences.

Sir John asked if God is perhaps "the only reality," as with the title of his book *Is God the Only Reality?* Such a large idea may be hard for some readers to appreciate or take seriously, but it has its place in many spiritual traditions in Christianity and beyond. Indeed, Sir John felt that in past centuries we were too simple-minded scientifically to understand these deeper truths and take them seriously, but accelerating discoveries in physics and cosmology now may help us recover them and make new advances in spiritual knowledge. In other words, there may come a time when science unveils what in the past was merely a matter of belief, or the self-reported experience of the mystics.

Is God's love the very mortar and matrix of the universe? In every instant of our lives are we sustained by a trustworthy Creator, however oblivious we might be to this? Is existence itself the perpetual triumph of love? As Sir John wrote on July 1, 1995, in an essay titled, "Like a Wave on the Ocean," "God may be the only reality—all else may be fleeting shadow and imagination from our very limited five senses acting on our tiny brain." What a contrarian idea, but those who have had those disorienting, awesome, and astonishing moments of spiritual experience in which time stands still, space disappears, and God's love is absorbed in all its *mysterium tremendum* know that reality as we perceive it is not the whole story at all.

Sir John surmised an Unlimited Love energy underlying all the laws, order, and being of the universe that "set up" (the "anthropic principle") evolutionary "complexification" and a creature capable of knowing God's love. He did not claim that this was more than a plausible hypothesis at this time from the scientific perspective:

> Certainly, there seems to be no conclusive argument for design and purpose, but there are strong evidences of ultimate reality more fundamental than the cosmos. So, if there are phenomenal universal forces, for example, in gravity, in the light spectrum, or in electromagnetism, can there not also be a tremendous unknown or non-researched potency or force of unlimited love? With earthly information now doubling every three years, can our comprehension of some

of these intangibles of spirit also be multiplied more than one hundredfold? Could unlimited love also be an aspect of dimensions beyond what we presently know as time and space? Could unlimited love be a universal concept beyond matter and energy as they are currently understood? To what realms beyond the physical might unlimited love reach? Just how vast is the reach of unlimited love?" (2003, 95)

We see here Sir John writing of Unlimited Love as something even beyond matter and energy, as somehow preceding these in the form of an infinite Mind that pronounced in a "big bang" something akin to the passage from Genesis, "Let there be light," or to the Gospel of John, "In the beginning was the Word."

Not every spiritual thinker and leader takes love to the level of Ultimate Reality like Sir John did, but many have. The great Hindu leader Mohandas K. Gandhi, in his classic work *The Law of Love* (1957/1970), refers to love as "more wonderful than electricity." "Scientists tell us," he writes, "that without the presence of the cohesive force amongst the atoms that comprise this globe of ours, it would crumble to pieces and we cease to exist; and even as there is cohesive force in blind matter, so must there be in all things animate; and the name for that cohesive force among animistic things is Love" (1957/1970, 5). And he went much further to write that "love sustains the earth" (1957/1970, 9). Here Gandhi was writing of love not just in what we *do* but in the very matrix of the universe. Gandhi's thoughts on love were clearly grounded in the same Hindu tradition that Sir John so greatly appreciated, in which God is explicitly characterized as Ultimate Reality.

The Quintessential Hypothesis

Here is Sir John's biggest question in the form of a hypothesis: *Ultimate Reality is a matrix of Unlimited Love (variously described by Sir John as the Absolute, the Supreme, the Ground of Being, a Higher Power, Infinite Intelligence, God, Godhead, etc.) underlying and constantly sustaining all the energy, matter, and mind in the universe and in ourselves.*

The method. We assess this hypothesis for plausibility, not proof certain, through (1) current social science of self-reported human experience of Unlimited Love, (2) current science of human joy and health in the context of love for God and neighbor as self, and (3) the evidence from the physical sciences, especially physics. If evidence from these three domains appears to more or less converge on our hypothesis, we have at least established its plausibility.

The caveat. Science may not develop for decades the methods to formally prove or disprove this hypothesis, although scientific progress is accelerating more and more rapidly, even exponentially.

PASTOR JOHN: THE BIG QUESTION IN RELATION TO THE SPIRITUAL PROBLEM OF EMPTINESS

The hypothesis above is not something that interested Sir John for purely intellectual reasons. He once said that had he not been a good stock investor, he would likely have been a minister. He did try his hand in midlife as a Sunday school teacher in New Jersey, just as he did in his Tennessee boyhood. There was always a pastoral aspect to his life and style of interaction, and he opened his board meetings with prayer. He wished to help us all on the unending journey toward greater meaning and purpose in life—a journey filled with desperate detours, destruction and self-destruction, and inevitable suffering. Hard lessons are learned hard. *Purpose is the question; Unlimited Love is the answer.*

The timeless and universal spiritual yearnings of human beings have always prompted questions about Ultimate Reality, love, and our significance. *Why am I alive? Does my life matter? What is my purpose?* The prophet Jeremiah asked, "Why was I born?" (Jeremiah 20:18). Are we alive just to navigate the struggle for survival, cope with sorrows and losses, and "succeed" while still feeling incomplete? Are we in the end no more significant in this universe than a warm bacterial soup? If there is no God of Unlimited Love, does all discussion of human purpose become obsolete and ridiculous, as the atheist philosopher Bertrand Russell honestly asserted?

In the Image of God

Does my life matter? Yes, because every human life matters to God, who shared a drop of Divine Mind and its immnse creative potential with each of us. Each individual, emphasized Sir John, is made *in the image of God* because we each have within us a small portion of the eternal Mind, which is our true self. The Jewish tradition speaks of God breathing into each of us an eternal God-connected spiritual element. If we are spiritually aware, we sense that God is active in us all, pushing our minds to grasp true ideas and carry them into creative expression consistent with love. *We are venues through which the Divine Mind can further express itself in our love and creativity.* In each of us, Divine Mind is completing another degree of unfoldment. Sir John wrote,

> The Bible tells us we are "made in the image of God." What does this mean? How would we describe our perception of the image and likeness of the Creator? Could this mean the divine presence resides as the deepest and most intimate spiritual reality within each human personality? Could this inner-spirit-spark radiate the Creator's unlimited love and creativity directly through each one of us? Can the sacred presence provide a source of spiritual guidance? (2003, 118)

Yes, we have deep within us, each one of us, the spark of divine creativity and love, channeled into so many expressions from fine arts to business to caring for a family (p. 118). Does my life rest on some cosmic mistake? Am I really no more than an organized collection of chemicals? Can my consciousness be explained simply in terms of chemical reactions in my brain? The answer for Sir John is no, because every life includes a little bit of God's eternal essence. Each life is therefore so significant that God wants to love each of us forever.

Our Purpose Is Love and Co-creativity

We are alive to experience joy, but joy flows not from self-love alone but from the love of self that is interwoven with sincere love for God and neighbor—the three conjoined in a tapestry. Within so many spiritual systems, deep joy comes from an awareness of God's surprising and overwhelming affirmation of our very being and that of our neighbors. Sir John had nothing at all against the love of self, but he endorsed only the right and most enduring self-love that comes when we discover the joy of abiding in the double love commandment (Matthew 22:37–40) that he often cited, as in the above passage.

Life is never a straight line, but it helps to have a noble purpose. Most of us have been anxiously concerned that someone we care about seems to lack even the hint of noble vision or elevating purpose. We see a vacuum of emptiness in his eyes and facial expression, and we hear it in his voice. Human beings need meaning and purpose to flourish. Sir John believed that most lives of noble purpose are shaped by the power of love, and by a loving God who lies in the very grain of the universe, in the very forces and energies that underlie and shape the visible world, and also within us in our minds. Our most noble purpose, he believed, is to live in the light of Unlimited Love and to increase human awareness of it (a) in every single one of our encounters and (b) worldwide both now and for future generations through spiritual progress.

Sir John clearly affirmed that each of us has a capacity to connect with Ultimate Reality:

> The true, universal self within us is an individualized center of God consciousness. As we become more willing to release the personal ego, we open up the door to greater communication with God. The one who relies on his own wisdom, beauty, skill, or money seldom relies on God. But the one who is humble and grateful for all such God-given blessings opens the door to heaven on earth here and now. (2012, 140)

Here Sir John emphasizes the humility that can invite each of us to be more deeply at one with a Higher Power.

Second, our purpose is to continue the work of creation as a co-creator. Each of us is given the precious mind that we have to be the hands, voices, and creativity of God. We each have the seeds of divine creativity within us, and our responsibility is to cultivate that seed and use it for loving purposes. Sir John felt that God gives us the spiritual part of our minds in order to be inspired with creativity and continue the work of Creation.

UNLIMITED LOVE DEFINED

In a letter from me to Sir John dated August 30, 2001, in which I had asked him for his definition of *Unlimited Love,* he faxed back a copy of my letter (August 31) with the following handwritten note on it:

> Dear Stephen,
> Congratulations on an excellent beginning! Hopefully you can craft a "definition" which avoids implying that "unlimited love" is only from humans toward humans. Possibly love is older and more universal than gravity. Possibly humanity is just the latest creation on one planet of the unlimited creative reality called LOVE? You may want to read carefully my little book "Pure Unlimited Love." God Bless you. John M Templeton, 31-8-01

I understood clearly that Sir John had in mind something much greater than mere human empathy, affection, and the like.

He faxed me another letter of response a day later, on September 1, 2001, just ten days before the attacks on the World Trade Center:

> A complete definition of "unlimited love" is not possible by humans, because perceptive abilities of humans are so limited. Instead we can research possible pathways toward unlimited love.
> Unlimited love may be billions of times more vast than any one temporary species on a single planet can yet comprehend. . . .

Gravity is easier to measure than love; but the power of love may be more creative, more timeless, more vast, more beneficial.

As we develop methods to increase our perceptions, we may discover more about unlimited love and the biblical words, "God is love; and he who dwells in love, dwells in God and God in him."

In short, Sir John was a bit hesitant to overdefine Unlimited Love because we do not know enough about it yet to be very precise. I often thought back to these exchanges after 9/11 and felt that human "spiritual progress" really did need to be accelerated because human beings distort the glory of God with so many sociopathic anthropomorphisms and then act accordingly.

However, Sir John did approve these words used on the website of the Institute for Research on Unlimited Love, which were distilled from his writing:

> The essence of love is to affectively affirm as well as to unselfishly delight in the well-being of others, and to engage in acts of care and service on their behalf; Unlimited Love extends this love to all others without exception in an enduring and constant way. Widely considered the highest form of virtue, Unlimited Love is often deemed a Creative Presence underlying and integral to all of reality: participation in Unlimited Love constitutes the fullest experience of spirituality.

Sir John liked the word "presence" and, having used it on page 34 of *Pure Unlimited Love* (2000b), he highlighted it for me several times. He meant a true living presence of Unlimited Love that is all around us even if we are oblivious to it. Now, this by no means is the precise wording Sir John would have articulated in the solitude of his study, but it seemed to include some of the elements he wanted, and it is certainly resonant with his many writings from which I tried to distill it. He definitely wanted the idea of a "Creative Presence underlying and integral to all of reality," because Sir John was less interested in love between people when this is separated from the love for and from God. So, any definition of Unlimited Love is bound to limp, but this one seemed serviceable enough for the nonce.

I did share another definition of love with Sir John that he thought was very clear and understandable. This came from the great University of Chicago psychiatrist Harry Stack Sullivan, who wrote, "When the happiness, security, and well-being of another person is as real or more real to you than your own, you love that person." There is truth in his definition. After all, whether we are looking down at a child sleeping, talking with a close friend, or listening attentively to someone's narrative of a hard life, goodwill and concern for well-being, security, and health seem to be the common thread of love. Sir John would merely add that God alone is capable of loving all people in this definitional sense, and that we are too when we abide spiritually in God's love.

Sir John was very clear, though, that he did not want to confuse Unlimited Love with altruism. On August 3, 2001, he wrote a letter in response to my query to him about the name of the institute. Our lawyers had already petitioned the state of Ohio for nonprofit corporate status (July 2001) using the title "Institute for Unlimited Love and Altruism Research." I brought this to Sir John's attention by fax, and he faxed back that it should be called *The Institute for Research on Unlimited Love*, because this was the only title that would allow a focus on God's love. He was entirely correct. So on September 7, 2001, we reported a name change to the state of Ohio and the IRS, advising them of the Institute's new name. We received our response of approval from the IRS on November 2, 2011. Sir John's August 3, 2001, letter included the following statement: "The somewhat related subject of altruism is important also," but "altruism could be included as one of the many consequences of unlimited love."

Sir John's hesitancy about "altruism" made sense to me. Altruism, or "other-regarding action," might be the result of an evolved innate rescue impulse, or of a moral principle ("a common humanity" [Kant]), or of habituated virtue and good modeling, or of empathy and compassion (see the Dalai Lama, Hume, Smith), or of meditation deflecting attention from self. But Sir John was especially interested in that form of "other-regard" that was grounded in spiritual experience, in alignment with a Higher Power, that had to do with the living reality of Unlimited Love.

"PURE" UNLIMITED LOVE

Sir John titled his most significant essay on this subject *Pure Unlimited Love*, knowing that human love alone can be tainted by the desire for reputational and reciprocal gains, manipulative and controlling tendencies, an exclusive focus on the nearest and dearest that demonizes outsiders, and self-destructive tendencies. We see hints at *pure* unselfish love in such things as benevolent friendships, parental love, and compassion for the neediest, but as anyone who picks up a newspaper can see, human love without some grounding in Unlimited Love is often unwise, narrow, inconsistent, impatient, harsh, and even easily inverted into rage and violence. The normal human being harbors a certain natural empathy, but this capacity is easily overwhelmed by poor role modeling, negative hierarchies (e.g., gangs), greed, selfishness, and the will to power. Sir John was a realist about human nature: "This [Unlimited Love] does not mean that you need to admire each person or weaken legal penalties for crimes. It does mean that if your mother were murdered, you should try to eliminate the poisons of hatred and revenge. While a murderer is being properly punished and prevented from a criminal life, *agape* love allows you to pray for his conversion and his soul" (1999, 1).

Writing this book is mind-expanding for me because Sir John wrote as much about deep physics and Ultimate Reality as he did about love and Unlimited Love. Bringing the two together requires an appreciation for his reflections on matter, energy and the laws of physics, divine idea, mind, spirit, humility, and many other themes. *Is Ultimate Reality Unlimited Love?* draws on the full range of Sir John's writings, many of which are posed as questions for exploration—for example, "Can love be an eternal universal force more potent than gravity, light, or electromagnetism?" (2000b, 32). Talk about a big question! Sir John challenges us to think about a big idea: Could Unlimited Love underlie all the laws of the universe, and all of being itself, including energy? Is all energy a modulation of a divine universal prime energy? Could this be the ineffable Ground of Being? This type of mysticism evokes the Sufi love poet Rumi, whom Sir John quoted as follows:

> O Love, O pure deep love,
> be here, be now.
> Be all; worlds dissolve
> into your stainless endless radiance,
> Frail living leaves burn with you
> brighter than cold stars:
> Make me your servant,
> your breath, your core. (1999, 6)

Is Unlimited Love like gravity? Sir John wondered if Unlimited Love is a force always present in our lives, however much we are utterly unaware of it. Perhaps God's love holds us close to it like we are held close to the earth by the energy of gravity—but of course we do not actually *see* gravity, nor are we conscious of it. Maybe much of the goodness that occurs in the world is more inspired by such energy than we realize, including all the astonishing synchronicities in life when we feel that someone said or did something that was in answer to a prayer. If Unlimited Love is like gravity, a basic force in the universe, then like gravity if it ever stopped we would know it instantly.

ULTIMATE REALITY DEFINED

Ultimate Reality as Sir John viewed it is something we have not yet understood and may never grasp fully. His hypothesis was that Ultimate Reality is divine creative thought constantly shaping and sustaining all that is, energized and motivated by a Godhead or Divine Mind of Unlimited Love.

Sir John saw a spiritual crisis in our age in which sensate materialism drives people away from a culture shaped by spiritual principles and new "bigger" ideas about Ultimate Reality. I believe that Sir John saw the culture of sensate materialism at the end of its rope, in a stage of decay and senility. He was clearly an ontological idealist in the sense that Ultimate Reality is not matter. He saw a more creative epoch on the horizon as the light of sensate materialism fades in the shadows and human beings become aware of who and what they are as beings uniquely participating in Divine Mind.

These ideas should not be surprising. How else would someone influenced by American transcendentalism, the Unity school of Christianity, the Cumberland Presbyterian Church, world religions such as Hinduism, and the ontologically idealist strands of modern quantum physics post-1930 be expected to think?

The word *ontology* means "knowledge of Being." Sir John's ideas lie broadly within the category of ontological idealism, which asserts that mind and spiritual principles are fundamental to the universe as a whole, as opposed to the view that mind and spiritual principles emerged from or are reducible to matter as epiphenomenal. Classical theism is an idealistic approach in that God, the uncreated and unconditioned Mind and Spirit, created everything *ex nihilo* (from nothing, other than divine prime energy) and is more fundamental than anything that was created. This outlook does not make the material world unreal or unimportant, but it does assert that the material world is an aspect of or manifestation of a deeper sustaining matrix of Divine Mind. In all idealist ontology (order of being), some ultimate spiritual reality exists beyond what appears to common sense and ordinary experience.

According to classical monistic idealism, matter is real but secondary to consciousness or Divine Mind, which is itself the grounding of all being. Consciousness and thought are fundamental. In the Judeo-Christian tradition, as *logos* proclaims, "Let there be light, and there was light." In Hinduism, all energy and matter emanate from the Brahman or Godhead. Monistic idealism is the antithesis of monistic materialism, for it makes consciousness, not matter, fundamental. Most spiritual traditions are historically grounded in monistic idealism, as are many philosophical traditions.

The American transcendentalist movement of Christianity was steeped in the ontology of monistic idealism, and was powerfully and definitively interwoven with the Unity school of Christianity and the New Thought movement. In one of his most famous poetic essays, "The Oversoul," the deeply Christian Emerson wrote words that echo into Sir John's thought world. The Over-Soul, "within which every man's particular being is contained and made one with all other," is very much something Sir John would endorse. Further, "We live in succession, in division, in parts, in particles. Meantime, within man is the soul of the whole; the wise silence; the universal beauty, to which every part and particle is equally related; the eternal ONE," for "the subject and the object are one." In this oneness with the divine One, we flourish: "When it breathes through his intellect, it is genius; when it breathes through his will, it is virtue; when it flows through his affection, it is love." Emerson continues, "We know that all spiritual being is in man," and "there is no screen or ceiling between our heads and the infinite heavens, so is there no bar or wall in the soul where man, the effect, ceases, and God, the cause, begins. The walls are taken away." We are the beneficiaries of "an influx of the Divine mind into our mind." This Over-Soul is in us, as we are a small piece of it, and in freedom we are able to create in love as the instrument of Divine Mind on earth.

To those versed in spiritual thought, history of religions, and the perennial philosophy (discussed later in this book), all of these ideas are unfamiliar. The main expression of mysticism in Judaism is the Kabbalah, which began to be taught in twelfth-century Europe. In Kabbalistic metaphysics, there are ten *Sephirot* (divine emanations) through which an infinite and incomprehensible God continuously creates and re-creates the material and spiritual dimensions of existence. There is a "light that fills all worlds" (divine immanence) as well as "light that surrounds all worlds" (divine transcendence). In the ancient Greek philosophy of Neoplatonism, the divine *nous* or Mind is that from which all of reality continuously emanates.

In Hinduism, again, Brahman is the universal Spirit or Mind that is the origin and the continuous sustaining support of the entire universe. Brahman is referred to as the Absolute, the Godhead, Ultimate Reality, and the Divine Ground of all being. The degree to which Brahman is personal and loving varies according to the school of Hinduism, but is especially strong in the tradition of *Bhakti*, with its doctrine of celestial love. All the great sages whose teachings are captured in the Hindu Upanishads teach that Brahman is the origin of all things, even of our own minds, such that an enlightened individual is aware that the true self (Atman) is a gift from and participation in Brahman. We are each then like a drop of water proceeding from an infinite pool of Water. Ultimately, our souls (Atman) or Minds are of the same eternal spiritual nature or "stuff" as the Brahman. Everything we see in the world and everything that is in the universe (material and spiritual) are manifestations of the different energies of the original and sustaining Absolute Mind, which contains within it all the archetypes behind all phenomenal forms. Brahman is filled with bliss, and to the extent that we realize our participation in it, we also experience bliss or ecstatic joy. Those who realize this are called *Brahmins*. Atman is the inner essence of the human being. *Atman* is the Sanskrit word that means "self," and one's true self is not part of this material world but instead is eternal, spiritual, and of the same essence as Brahman.

Kalam is Islamic systematic theology, and it encompasses a number of schools. The dominant one, for centuries now, has been Ash'arism, which opposed and overcame the famous Mu'tazilite rationalist school of theology. Ash'arism expounds the doctrine of "occasionalism," that God re-creates the world at every instant, essentially in the way that God sees fit (leaving the door open for miracles), though God almost invariably follows "habits" (allowing for the formulation of laws in nature). Ash'arism doesn't say that the first principle that God acts upon is love, but the latter is certainly one of God's attributes and defining features.

Sufism, however, another branch of Islamic thought, puts more

emphasis on love as the defining relation between God and God's creatures, including humans. Sufism is a general term for Muslim mysticism
as it arose in response to the worldliness and materialistic values of
Islamic leaders in the eighth century. Teaching consciousness or awareness of Divine Presence and pure love, Sufis practice nonviolence and
understand God as an Ultimate of perfect universal love. Sufism pro-
fesses that God is the All-Encompassing, and in some bolder versions,
God is the All-That-There-Is, so in essence we swim in God's manifestation; to be more accurate, we and the world at any instant represent
the state that God wanted to be in or wanted to manifest. A Sufi believes
that our goal should be to feel God and identify with God, which is the
purpose of worship. So in this school, God is definitely the Ultimate
Reality, even the Only Reality, and God is Unlimited Love. One of the
mainstays of Islamic thought, at least in some schools, is the idea that an
immanent God is continually re-creating the universe and all things in it
at every instant through Divine Love. Thus, every moment represents a
new universe formed in the energy matrix of love. The book sitting on
your desk looks as it appears, but at the deepest energetic level it is in
constant re-creation. This sounds odd, perhaps, but it is not too much
different than what comes from the realm of quantum physics. So it is
that Sir John Templeton quoted the Sufi poet Rumi with appreciation,
and invited the Islamic scholar Sayyed Nasr to his home for discussions
of Unlimited Love in the Sufi tradition.

According to William C. Chittick, the authoritative Western scholar
on Sufism and love in the Islamic tradition, the philosopher Avicenna
in his *Treatise on Love* explains that love motivates the Absolute Good to
create the universe (2010, 177). Those who "actualize their full potential
to love God and to be loved by Him" are sometimes called "perfect
human beings" (2010, 180). Moreover, "Despite the emphasis in most
texts on earning God's love by following the Prophet, many authors
stress God's unconditional love" (2010, 182). As Chittick sums up
Islamic thought, "Love is the very Reality of God Himself. It gives rise
to the universe and permeates all of creation. God singled out human
beings for special love by creating them in His own form and bestowing

on them the unique capacity to recognize Him in Himself and to love Him for Himself, not for any specific blessing" (2010, 193).

Religions refer to the nature of God or Ultimate Reality in different ways, as a personal loving God and as an impersonal Being. In Judaism, Christianity, Islam, Sikhism, and certain traditions within Hinduism, God is highly personal, loving each person in a deeply personal way. But there is also the image of Ultimate Reality as impersonal Being—the Primal Tao of Chinese Taoism, the Aristotelian Unmoved Mover, the "Suchness" of Mahayana Buddhism, and the Brahman of Hinduism at least in some of its traditions. There is also the idea that Ultimate Reality is immanent within each of us, that we each are given forever a small inner light or soul that is in essence a piece of the eternal Mind of God.

It cannot be overemphasized that *Ultimate Reality* with regard to the history of religions is a particular translation of the Hindu term Brahman. It is somewhat different than the Abrahamic concept of God in that it focuses attention on the deepest level or underlying matrix of all of reality, from which everything in the universe emanates. God is not "above" reality but behind it. In the Hindu tradition, this Ultimate Reality is Divine Mind or Infinite Thought, and this Thought can manifest itself in the energies, matter, and physical laws of the universe. Sir John was a Christian, but he was fascinated by various other religions, especially with regard to their conceptualizations of God and love. As Glenn R. Mosley underscores, Charles and Myrtle Fillmore studied Hinduism with great diligence, and their emphasis on God as Ultimate Reality had roots in this tradition, which was so popular in the spiritual circles of their transcendentalist culture (2006, 11). Sir John lies squarely within the *philosophia perennial*, the perennial philosophy, which has its deeply spiritual element. In the words of one interpreter, this is "the metaphysic that recognizes a divine Reality substantial to the world of things and lives and minds; the psychology that finds in the soul something similar to, or even identical with, divine Reality; the ethics that places man's final end in the knowledge of the immanent and transcendent Ground of all being" (Huxley 1944/2009, vii). What we shall discover is that Sir John, consistent with the mystical perennial philosophy of old but carried forward through the Unity school of Christianity, surmised that our eternal mind or "soul" is "identical with, or at least akin to, the divine Ground" (Huxley 1944/2009, 1).

So here I offer an interpretive description of Sir John's primary thoughts on Ultimate Reality:

- Ultimate Reality is the deepest substrate or underlying energy matrix of the universe that lies behind the virtual reality that we see.
- Ultimate Reality, while everywhere and in everything as the Infinite Intelligence from which all things emanate (but that includes within it original energy), is present in a special form as the "eternal part" of each human mind (soul), which Sir John wrote of as in essence a "little part of God."
- Ultimate Reality, while often considered impersonal in Hindu thought, in this Christian synthesis is equivalent to Unlimited Love (Sir John referred to "God is love" as the greatest theological "equation"); moreover, because we have a soul that connects us with this Ultimate Reality, we are able to receive such Unlimited Love and participate in it, thereby rising to a level of universal and unconditional love of which unaided human nature is simply incapable.
- Ultimate Reality is infinitely creative. By virtue of having souls of the same essence, we also are capable of astonishing creativity that is frequently self-reported as involving God's inspiration; moreover, as co-creators with God we should always pursue loving ends.
- Our purpose in life is to experience joy (bliss) through awareness of God's love for each of us, our love for God, and our love of neighbor as ourselves.

This Ultimate Reality is a fountainhead of Pure Unlimited Love. In *Story of a Clam*, for example, he noted,

> Agape!
> Feel the essence of pure love glowing ever more brightly within you, bringing forth warmth. This love can transmute the last residue of old hurts, misconceptions, and egotistical

ignorance into greater awareness and understanding. It can help you release any emotional remnants of guilt or shame. And this love can invite the larger warmth of the Great Creator's Unlimited Love to dwell within you eternally. (2001, 91–92)

To Sir John, "this essence of pure love glowing ever more brightly within you" is something we can come to become increasingly aware of or "feel." It is something that seems to already be there within us, but we are oblivious to it. The tone of this passage is more meditative or contemplative, more one of realization than of a singular dramatic episode. As awareness grows, destructive emotions and thoughts are "crowded out"—a favorite terms of Sir John's. Then there is room for "greater awareness and understanding" of Unlimited Love "eternally." We must take seriously the phrase in this passage, "dwell within you eternally." Across Sir John's many writings, growth in love is our purpose on earth—preparing for the eternal life of that aspect of the human mind that he referred to as "the soul," because it is quite literally a piece of the Infinite Mind given unto us in love as our own, desiring reunion with its Source.

Chapter 3

Sir John's Humble Approach to Ultimate Reality

SIR JOHN FELT that God's love for all people without exception, and our transforming experience of it, is the most important point of convergence between all significant spiritualities and religions. We marvel at the ways and power of this awesome love, considering it the best hope for a far better human future. How do our complex brains, unique imaginations, communicative abilities, reasoning powers, moral sense, and—most of all for our purposes—their spiritual promptings and sense of solidarity with Divine Mind and Unlimited Love give rise to unconditional, universal, and unchanging love for our neighbors or for those we do not even know? Herein lies the key to all spiritual and religious progress, and hopefully we will see such progress before the great religions—all of which are institutions created by flawed human beings and therefore capable of destructive actions—contribute to annihilation in the arrogance that Sir John so feared. Sir John clearly saw science as holding the key to spiritual progress in this area, and as holding the ultimate key to greater human awareness of a love that is infinitely greater than mere human love, a love that is truly Unlimited Love.

Sir John's appreciation for progress in religion through scientific discovery and a thorough understanding of the theological implications of such discovery was deeply shaped by the influences of Unity, which eschewed theological stagnation, stressed how much we have yet to learn, and looked forward to an astonishing human future with light millennial intonations.

SCIENTIFIC INVESTIGATION SHAPING INSIGHT INTO UNLIMITED LOVE

Sir John makes clear in one of his earliest books, *The Humble Approach: Scientists Discover God* (1981/1995), that we need to discover more about a love that is both "unconditional" (free from reciprocal expectations or assessment of properties deemed desirous in the recipient) and "universal" (equal-regarding in its concern for all people without exception), for this defines the essence of the Divine Mind in which we each participate to varying degrees and with varying awareness. Sir John commonly used both of these terms together to describe a *fully extensive and freely bestowed love.*

Nothing thrilled him more than the potential of progressive science to bring greater clarity to the perennial laws of nature and spirit that are captured in scripture and theology and have universal significance. This approach would actually "keep theology in the vanguard of the knowledge explosion" (1981/1995, 66). Sir John made the astonishing suggestion that churches and foundations should support "joint theological-scientific research" (p. 67). I emphasize his word "joint," for Sir John was interested in a scientism explicitly in dialogue with spirituality and theology. He further noted, "More and more manpower and resources are being devoted to the forces of nature—discovering, proving, understanding, using and teaching these forces. But almost everyone agrees that one of the greatest forces on earth is *love*. Should churches finance constant research into this force of love?" (1981/1995, 124). We should not infer that churches should do so exclusively or to the detriment of many other important functions.

Why a new, more scientifically shaped approach to the study of Unlimited Love? Sir John could see in his lifetime the immense progress being made in so many domains because of science and technology. Ours is an age of ingenuity. He often noted that more progress has been made in the last century than in all the centuries preceding it, and progress is accelerating. Sir John greatly admired Thomas Edison for all of his transformative inventions and for focusing his resilient creative energies on projects that he felt would be of the greatest benefit to humanity. But with progress in science and technology "speeding up," as Sir John put it, we should be able to make equally significant advances in the spiritual assets of life, such as love, forgiveness, creativity, gratitude, Ultimate Reality, awe, self-control, and the like.

Spiritual progress is certainly a dominant theme across all of Sir John's many visionary writings, and spiritual progress in particular requires a humble awareness of how little we know in order for us to be truly "eager to learn":

> Many religious people are not yet inspired to hope that the spiritual future could, or should, be improved from anything that has ever been learned before. Many do not imagine that progress in religion may be possible, perhaps by appreciating ways that sciences have learned to flourish and by being creatively open to a discovery-seeking and future-oriented perspective. For so many religious people, the future of religions seems nothing much beyond the preservation of ancient traditions. Some therefore may not want to consider the possibility of a future of progressively unfolding spiritual discoveries. Yet, if our creativity has significance, then what Newton called the "ocean of truth" may hold wonderful possibilities for the future of religion, as it clearly does for the future of science.

Could the adventure of science both inspire and assist religion to explore a rich future of "boundless possibilities"? (2000a, 5–6). Such a wide-open vision of spiritual progress can only succeed when religious people affirm how much there still is to learn about the mysteries of spiritual principles and realities.

Indeed, the "humility theology" or the "humble approach" that Sir John developed rejects parochial close-mindedness and a too-easy contentment with the current state of spiritual understanding. One of his most succinct statements of the "humble approach" is this:

The humble approach is meant to help us as a corrective to parochialism in religion. Humility reminds us that our concepts of God, the universe, and even our own selves may be too limited. It is universally a wise teaching in most great religions that we are all too self-centered. We overestimate the small amount of knowledge we possess. To be humble then means to admit the infinity of creation and the boundless possibilities within it. Thus can we become motivated and enthusiastic to search for opportunities for us to engage in creativity and gain benefits from experiences and possibilities we may never have dreamed of before? Such an approach asks each of us, whether we are scientists or mainly active in religious-focused lives, to become open to the possible abundance of spiritual potential in our own lives. (2000a, 6)

Sir John wanted us to "cultivate a spirit of humility simply by being open to the possibility of our existence within a divine reality which dwarfs our personal reality" (2000a, 6). He felt that with humility we can "participate in this accelerating creative process." (2000a, 7).

Less Than 1 Percent

Sir John's sense of humility had roots in the Presbyterian tradition's emphasis on the glory of God and our incredibly small powers to even begin to grasp divine intelligence and the potential of Unlimited Love:

By the word *humility* we are not referring to self-denigration but instead to recognition that no human has yet found even 1% of those basic invisible realities called spiritual, such as love, purpose, consciousness, intellect, creativity, worship and thanksgiving. (2000a, 13)

Putting a symbolic percentage on what we know now and what we do not know (99 percent plus) is by no means meant to be taken literally.

The point is that "Every person's concept of god is too small" (2000a, 7). In such provocative terms, Sir John challenged us all with the idea of humility, as human pride is the great obstacle to spiritual progress:

> Through humility we can avoid the sins of pride and intolerance and avoid especially harmful religious strife because it is unlikely that any religion could know more than a tiny bit about an infinite god. Humility opens the door to being hungry to discover basic realities of the spirit. (2000a, 8)

Here Sir John was concerned with the potential of religious arrogance to cause violence. He saw arrogance as a problem across many faith traditions, not just one. After all, "If any person were to say he knew all about god, it might be like a tree claiming to know all about the gardener" (2000a, 17). He was concerned about terrorism well before 2001, referring to it as the major obstacle to world economic growth.

Sir John lamented the fact that many educated and scientifically minded individuals have no interest in religion, and he saw "the humble approach" as a way of engaging them:

> Many highly educated people feel that religion is obsolete. In some senses they have a point. We typically do not observe the kind of dynamism in religion that we see in other areas of life such as science, technology and business. To many, religion sometimes seems like a kind of history museum which lacks the excitement and vibrancy of other aspects of life that constantly experience innovation. (2000a, 9)

So did Sir John analyze the agnostic mindset (no one, not even Richard Dawkins, claims that God can be entirely disproven, and therefore honest atheists are really agnostics or else they are making statements of their own faith that there is no God) and with some considerable understanding. He certainly did not react to such views.

Renewing the Spiritual Traditions

Sir John always affirmed the value of conserving great traditions of worship and spirituality. In doing so, he was aligning himself with ideas associated with Edmund Burke or Russell Kirk—that these traditions have evolved within cultures over centuries for good reason and confer a variety of selective advantages that should be respected:

> Could young people and intellectuals be attracted to forms of religion that are genuinely dynamic and rapidly progressing? Does this mean that the old ways have to be discarded totally? I hope not, because it is clear that much of the strength of religion is in the precious core of wisdom and truth that it transmits from one generation to the next. So opening up a few religious communities to new concepts and new adventures of spiritual learning should not be like a revolution which attempts to build upon the ashes of the old. *My own hopes for rescuing various religions from obsolescence would be for the visions and teachings of the great prophets of the past not to be disputed. Rather they should be studied again and considered together with recent concepts of reality as springboards toward creating new and even expanded understanding of divinity and inspiration in worship and ritual.* (2000a, 9, italics added)

Here Sir John is indicating his great appreciation for the evolution and role of traditions of spirituality, religion, and worship over time in forming a foundation for culture, civility, and social cohesion. He was an evolutionist rather than a revolutionist. Remember that in his teenage years Sir John was the superintendent of the Sunday school in the Cumberland Presbyterian Church in Winchester, Tennessee, and in his adult life he was a Sunday school teacher in Englewood, New Jersey.

Over all his years Sir John hoped for a renewed appreciation for an energized theology. After all, "Theology was once considered the queen of the sciences" (2000a, 10). Moreover, "It may someday regain that title, but first we need to learn how to learn in order to regain that title" (2000a, 11). The purpose of the John Templeton Foundation was stated thus:

> A major aim of the Templeton Foundation is to help those relatively rare and visionary entrepreneurs who are trying to encourage all religions to become enthusiastic about concepts of spiritual progress and new spiritual information, especially by linking with scientific methods and lines of inquiry. If benefits from this approach can be practically demonstrated, then it may be welcomed and can help reinvigorate appreciation for and to supplement the wonderful ancient scriptures which stand at the core of most religious cultures. (2000a, 11)

In addition to these "visionary" souls who are trying to reinvigorate theology in a new paradigm, Sir John also saw value in supporting "well-regarded scientists on basic areas with theological relevance and potential, such as love, prayer, purpose, altruism, creativity, thanksgiving and worship" (2000a, 11). May those who follow Sir John embrace the visionaries as he did.

With his visionary sense of urgency, Sir John did not wish to approach the writing of *Is Unlimited Love Ultimate Reality?* as would the typical divinity school professor or philosopher (analytic or phenomenological) who devotes years to the hermeneutical task of proper exegesis and analysis of theological and philosophical texts. He appreciated such scholarship, of course, but only to a degree. In the modern era, when science holds a dominant place in thought and culture, this scholarship, no matter how meticulous, could not have a wide influence. Sir John, who was especially interested in nurturing a new culture of deep spiritual wisdom, did not see possibilities for major "progress in religion" or "spiritual progress" as long as theology remains detached from the tremendously exciting domain of scientific advance. Sir John lamented that theology, which could be made exciting to the modern mind when deeply informed by leading-edge scientific advance, has become increasingly relegated to the archaic and culturally irrelevant.

He found such excitement in the theologians who were also physicists, the mathematicians who were also mystics, and the paleontologists who were willing to ask cosmological questions. In this dynamic dialogue his mind would overflow with an eagerness to learn, and he would take pen in hand either alone or with a trusted collaborator. Nothing excited him more than a new discovery that was ripe with implications for how we can rethink God and enlarge our conceptualization. Again, as he always wrote, "Everyone's God is too small."

No doubt Sir John was in the broad sense resonant with that common human yearning to experience the divine. In his writings we find ample reference to mystics like Rumi, Meister Eckhart, Fillmore, Tillich, Paul, the writer of John's Gospel, and a host of others united by a common intensity of spiritual thirst. Sir John was not attracted to dogma or any form of fundamentalism. Rather, he was captivated by those great minds who had a lively spiritual sense of divine reality being closer to us than we know—in us and with us rather than out there or above us—and about which we know so very little as of yet but with the help of science can hope over time to discover so very much more. Have you ever felt that you misplaced something that was right there in front of you or in your pocket all along? Suddenly you see it and cannot imagine how you lost sight of it because all along it was right before your eyes? Sir John felt that God is right here with us in every particle of reality, available to our minds, and holding us together like gravity holds together the universe—only we have lost that awareness.

Sir John envisioned a new era in which unhampered science would sit alongside theology or theophilosophy to shape a wonderfully exciting progressive conversation about the meaning and purpose of life, the shape of Ultimate Reality, and the place of love in the universe. While he believed that science must always move forward with its most objective methods and remain absolutely free of even the slightest theological, philosophical, or ideological influences, the advances made should all be taken up centrally for discussion in the domain of spiritual reflection so that ancient insights of perennial philosophy might be brought up to date and made so vibrant that no one would wish to ignore them.

The truly great theologians of Unlimited Love have always adapted their thinking to the science of their day, never seeing even the iota of a need for a "war" between science and theology. On the contrary, scientific findings have often been the fertilizer for theological development. Sir John always devoted a great deal of attention to the spiritual and theological implications of scientific findings, which he referred to as "dimensionality." Theology needs science to be the queen of the sciences.

This approach might quicken and enliven the spiritual traditions in which young people have lost interest, by freeing those traditions from the fundamentalism that eschews science as enemy. Sir John wanted to draw younger people into renewed conversation with the big spiritual questions of life through innovative science that enlivens an otherwise static theology. He was the teacher always helping younger generations to find renewed value in spirituality and faith. This could happen, he believed, only when the best of exciting leading-edge science serves as a theological lens. He wrote,

> My own hopes for rescuing various religions from obsolescence would be for the visions and the teachings of the great prophets and teachers of the past not to be disputed. Rather they should be studied again and considered together with recent concepts of reality as springboards toward creating new and even expanded understanding of divinity and inspiration in worship and ritual. (2000a, 9)

In other words, we need "recent concepts of reality" and scientific discoveries that can enrich and enliven ancient insights, rescuing religion from "obsolescence" by nurturing new insights into spiritual realities like Unlimited Love that should be "studied again."

The Rejection of NOMA
Sir John was not content to leave religious belief in the domain of human subjectivity and meaning. He was clearly not in resonance with

Stephen Jay Gould's idea of "NOMA" or "non-overlapping magisterial," where a magisterium is "a domain where one form of teaching holds the appropriate tools for meaningful discourse and resolution" (Gould 2002). According to the theory of NOMA, science covers the empirical magisterium, and religion covers questions of ultimate meaning and moral values. Gould, gentleman that he was, suggested that this division of labor is a consensus of people of good will in both domains. However, NOMA fails because scientists make all sorts of claims about how new advances put religion out of business, and the religions often claim miracles that impinge on the domain of science. Moreover, for Sir John, relegating spiritual progress to the domain of nonscientific investigation was anathema. After all, the entire first book of Calvin's *Institutes of the Christian Religion* is devoted to how the glory of God is revealed in and through the natural world, and most great scientists in the age of Enlightenment and discovery were viewing everything from physics to physiology as revealing the intelligence of God. The idea of separating spiritual progress from science was for Sir John the death knell for such progress, leaving religion buried in the past and consigned to increasing irrelevancy.

Rather, in the tradition of Presbyterianism and all the offshoots of Calvinism, as well as of Unity, Sir John was convinced that we could learn much about the Creator by observing the universe, which is a reflection of divine creativity (Romans 1:20). Indeed, many physicists, like Paul C. Davies, and some biologists, like Francis Collins, have written highly successful books about how science alone brought them to their knees in awe and humility as they came to the conclusion that there really is a God behind nature or a Wisdom underlying the mathematical laws of thermodynamics. Some astonishing mathematicians have concluded, as did Plato, that a higher reality underlies the equations of our universe. Sir John hoped that the very best young scientific minds would be part of a progressive and humble spiritual quest, rather than being caught up in the detours of fundamentalist belief systems that really are antiscience and antiprogress. One of the virtues that Sir John most espoused is a respectful open-mindedness. He believed that

God's greatest gift to each of us is our mind, and Sir John wanted us to use it.

PLAUSIBILITY IS IMPORTANT EVEN THOUGH IT IS NOT FULL PROOF

Humility is evident in Sir John's whole approach to knowledge. He raised such big questions; he speculated; he spoke of the plausibility of an idea knowing that actual certainty and proof as a scientific matter might or might not follow, or even be possible with current scientific methods. He was eager to learn but emphasized how little we know.

Late in his life he titled an important book as follows: *Possibilities for Over One Hundredfold More Spiritual Information: The Humble Approach in Theology and Science* (2000). In this volume he posed hundreds of questions for future investigation. He was often more comfortable raising a provocative question than asserting an answer, although a set of core beliefs clearly informed his approach. Thus in all the pages that follow, the reader needs to know that Sir John is not claiming certainty on matters of Ultimate Reality, yet he certainly sensed a basic direction in physics and cosmology that make statements about Ultimate Reality increasingly plausible. The big bang theory, the universal constants (speed of light, Planck's constant, the gravitational constant, the weak- and strong-force constants, and the like) pointed him to an Infinite Intelligence. Here he was in accord with great physicists like Fred Hoyle, who changed his mind openly from atheism to acceptance of a "super-Intellect" as the most likely cause of the universe. Big bang cosmology and the improbability of these elaborate, fine-tuned laws and constants coming out of an unintelligent beginning have convinced some great contemporary physicists—such as Arno Penzias, Paul Davies, Brandon Carter, Roger Penrose, Owen Gingerich, and many others—that some kind of Infinite Mind is plausible, or even necessary. The extreme improbability of the random occurrence of an event like a big bang through which a universe with energy, time, and space suddenly burst into being from nothingness—and the equally great improbability of universal constants (not to mention space and time) coming out of no originating intelligence—make the rationality of atheism tenuous. Stephen Hawking, an agnostic, nevertheless finds himself asking, "What is it that breathes fire into equations and makes a universe for them to describe?" (1988, 174). For Sir John, it was immensely exciting that first-rate physicists are brought by science alone to the idea of an Infinite Mind lying behind the "beginning" of the universe and its immutable laws, including their sustainability over time. He followed this development with great enthusiasm.

These days, physicists sometimes quote Augustine, who claimed there was no "time" before the beginning, the big bang. Sir John would join in asserting the plausibility of an Infinite Intelligence, since it appears to be a very reasonable position. And if this position is plausible, then why not at least ponder the question of why this plausible Infinite Intelligence would bring this astonishing universe into being? So it is that three years after *Possibilities*, Sir John wrote (with Unity minister Rebekah Alexander Dunlap), *Why Are We Created? Increasing Our Understanding of Humanity's Purpose on Earth* (2003) at the age of ninety-one. For Sir John—and for many spiritual people across the globe—the question points in the direction of Unlimited Love. (All of this, by the way, is logically and totally unrelated to any and all sociopathic images of God, and to any and all sociopathic activities perpetrated in God's name.)

The Most "Famous Equation"

Could God actually be constituted by love? In the mid-1980s, James Ellison undertook extensive interviews with Sir John and discovered that his favorite biblical quotation was 1 John 4:7–12, which includes, "Beloved, let us love one another; for God is love, and he who loves is born of God and knows God" (1987, 149). This is not surprising, for the Gospel of John presents a spirituality and practice of love that define true faith as well as the inner essence of "religion." This "binding together" (re-*ligeo*) of God, self, and other in triadic love is what really matters. The externalities of religions are of value, but only insofar as they contribute to this movement of love as an emotional and spiritual energy, and do not cause stagnation. Sir John wrote that the words "God is love" constitute the most "famous equation in theology" (2000a, 88). Whenever we feel and think that the security, well-being, and joy of another are as real and meaningful to us as our own, we love that person and we may be participating in God's mysterious love energy. Imagine describing a statement such as "God is love" as an "equation in theology." Sir John suggested that, someday, perhaps in several centuries, physics, chemistry, and mathematics might conceivably substantiate this equation. He asserted this with plausibility, not proof.

TRANSLATING SIR JOHN'S UNLIMITED LOVE VISION INTO THE PUBLIC WORLD AND CULTURE

We must always remember that Sir John wanted most of all to transform culture in the direction of spiritual progress. Sir John ultimately wanted to impact culture in a transformative way by translating breakthrough science into the everyday world of everyday lives.

Let me return to that conference with Sir John in October 1999, the perfect example of Sir John's approach. We who organized the conference settled on calling it "Empathy, Altruism, and Agape: Perspectives on Love in Science and Religion—A Research Symposium." We wanted to accommodate the scientific language of "empathy" and "altruism" as well as the more spiritual language of "love," and so this title seemed reasonably fitting, although we debated it heavily. The conference was convened at the University Park Hotel at MIT in Cambridge, Massachusetts, and lasted three days, from October 1 to October 3. The conference's goal was to initiate creative thinking toward stimulating and promoting excellence in research into the phenomenon and interpretation of self-giving love. The approach was highly integrative, linking the biological and social sciences with philosophical, ethical, and spiritual-theological themes. As the brochure read, "Many religious traditions affirm in various and diverse ways that love is the 'heart of being' and that the ultimate reality and ultimate purpose of things is related to love." We asked a range of questions—many of them the wrong questions in Sir John's mind because we were much too centered on mere frail human nature and human love. It was an exciting conference, but one that I eventually came to see, with Sir John's prompting, as far too limited. In the introduction to the brochure, we asked:

- To what extent do human individuals and societies manifest behavior that is motivationally or consequentially altruistic?
- What are the evolutionary origins and neurologic substrates for altruistic behavior?
- What developmental processes foster or hinder altruistic attitudes and behavior in various stages of life from early childhood onwards?
- What psychological, social, and cultural factors influence altruism and caring?
- How do spiritual and religious experiences, beliefs, and practices influence altruistic attitudes and behavior?
- How does the giving and receiving of altruistic love interact with personal well-being and health?
- How can researchers from various disciplines collaborate to enhance this field of study?
- Overall, is it possible to gain new insights which can be utilized to help people and their communities to better appreciate the significance and importance of love, and benefits from its expression as a lived reality?

What was Sir John's response? He was not overwhelmed because we were not adequately focused on divine Unlimited Love as a spiritual reality. He determined to write his booklet, titled *Agape Love: A Tradition Found in Eight World Religions* (1999), covering Judaism, Christianity, Islam, Hinduism, Buddhism, Taoism, Confucianism, and Native American Spirituality. Two years later, he wrote *Pure Unlimited Love*, in part as a reaction to our lack of focus on Divine Mind and Divine Love. No wonder he sent me all those letters in 2001 about taking as central the Unlimited Love that made humans, and not just to concentrate on human love. I knew that I had failed him a bit because I was being too human-centered.

We had convened forty of the truly premier leading panelists around these themes, half of them from the sciences (such as Dan Batson, Antonio Damasio, Frans B. M. de Waal, William H. Durham, Gregory Fricchione, Jerome Kagan, Thomas R. Insel, Samuel and Pearl Oliner, V. S. Ramachandran, Jeffrey P. Schloss, Lynn G. Underwood, and David Sloan Wilson) and half from spiritual, theological, or philosophical perspectives (such as Don Browning, Ruben Habito, William Hurlbut, Gordon D. Kaufman, Stephen J. Pope, Rev. Eugene F. Rivers, Dame Cicely Saunders, Elliot Sober, Lawrence Sullivan, and Edith Wyschogrod). We also had more than fifty attendees who on the whole were as highly regarded as the panelists. But we missed the point for the most part because we did not bring Unlimited Love, Godly Love, front and center. It was only on the margins. Hard lessons are learned hard. What had we really achieved? Not much.

The conference was a challenge because such highly regarded researchers and thinkers tend to be fairly absorbed in their own silos and niches, not necessarily recognizing fully how much progress can be made when they move out of those comfort zones into a wider framework. What really brought everyone together were the presentations by world-class exemplars of self-giving love, like Dame Cicely Saunders, the founder of the hospice movement, and representatives of L'Arche and its founder, Jean Vanier.

I recall spending much time with Sir John that weekend. We had several long conversations over meals, and time to talk quietly about what we were hearing and seeing. He was on the one hand so pleased with it all, but several times he returned to ask me, "Stephen, but where is God's love in all this? God's love is a million times greater than anything we are discussing." That was quite a statement from a man who never was critical of anyone. He knew that we were just forming a network and had a long way to go to make real spiritual progress. Two years later, with support from the John Templeton Foundation and building on the book that came from the MIT conference, I was honored to found the Institute for Research on Unlimited Love (www.unlimitedloveinstitute. com) in Cleveland as an independent research organization. Sir John stuck with me, but I knew that eventually I would need to grow in spiritual boldness and ask the really big metaphysical question of Ultimate Reality as Divine Mind and Unlimited Love. Sir John was patient, as though he knew that this was challenging.

The Institute began early one morning in June 2001 with a fax from Sir John about the idea of the institute. I sat down with a nonprofit lawyer in the Caribou Coffee House on Coventry Road in Cleveland Heights, Ohio, to found an institute for high-level research on love. Not any kind of love, mind you. Not giddy romantic love or love of chocolate or designer shoes. But the kind of love that affirms and finds joy in the flourishing of others, engages in acts of care and service on their behalf, and is both *unconditional* and *universal*. This kind of love—Unlimited Love—is exhorted as the highest goal of moral and spiritual awareness. The Institute for Research on Unlimited Love was incorporated as a nonprofit research entity a few months later and is now the world's leading scientific institute focused on the expansion of our understanding of Unlimited Love.

The institute's mission at inception was too broad, secular, and humanistic. Only a decade later did we have it right: *We seek to investigate the lived experience and Ultimate Reality of Unlimited Love through the high-level science (e.g., cosmology, mathematics, neuroscience, psychology, sociology and the health sciences) in dialogue with theophilosophical thought in order to contribute to spiritual progress.*

Our guiding questions, which finally came to be shaped by my readings of Sir John's books and some conversation with him (better late than never) became these:

- Does the self-reported spiritual experience of Unlimited Love result in increased benevolence and compassion?
- Do physics and cosmology point toward the same "Ground of Being" that those who report having experienced it often describe in terms of unconditional and universal love?
- How can the major spiritualities and religions of the world elevate and adhere to their various conceptualizations of Unlimited Love, and thereby bring out the best in their adherents?

- Does the love of neighbor contribute to the happiness, health, and resilience of those who give it, in addition to those who receive it?
- How can we raise children of Unlimited Love?

In one wonderful week (May 31–June 5, 2003), following Sir John's approach of working at the interface of science and spirituality, the institute convened a major international conference titled "Works of Love: Scientific and Religious Perspectives on Altruism—An International, Interfaith, and Interdisciplinary Conference" at Villanova University, working closely with Dr. William Grassie and the Metanexus Institute. We began the conference brochure with these words:

> Unselfish love for all humanity is the most important point of convergence shared by the world's great spiritual traditions. We marvel at the ways and power of love and find in it the best hope for a far better human future. People from all walks of life, often those disadvantaged themselves, excel in loving kindness, not just for their nearest and dearest, but also as volunteers and advocates on behalf of the stranger. How do our complex spiritual and religious promptings give rise to this remarkable yet not at all uncommon practice of unselfish love for our neighbors, or for those we do not even know? If we could answer this question and harness the extraordinary power of love, the world might well erupt into hope.

This was close, but not close enough. But the big questions that we sought papers on were fairly in line with Sir John's vision, so it was a better conference than the one at MIT in 1999. The topics and related questions and prompts were as follows:

- **Faith Active in Love: Lives of Compassionate Love**—Scientific and theological interpretations of remarkable lives exemplifying love for all humanity.
- **Unlimited Love and Human Nature: The Problem of Inter-Group Conflict**—To what extent is the substrate of human nature

receptive to the ideal of love for all humanity? Drawing on evolutionary biology, how does human nature resist this ideal?

- **The Healing Power of Love: Healthful Compassion**—To what extent does the giving of compassionate love and generous service impact the agent with regard to morbidity and mortality? How does this impact the recipient's health and therapeutic efficacy?

- **The Biology of Compassionate Love: Evolutionary, Physiological, and Neurological Correlates**—How can compassionate love be understood in the light of neurology, endocrinology, genetics, and immunology?

- **The Psychology of Compassionate Love**—What psychology models help to explain the experience of compassionate love and help to explain its key features?

- **The Sociological Study of Faith-Based Communities and Their Activities in Relation to the Spiritual Ideal of Unlimited Love**—Are volunteerism, social capital, and civility enhanced by faith in divine love? Are faith-based organizations more effective than secular ones in enhancing and sustaining "works of love"?

- **The Emergence and Impact of Helping Behavior in Young People**—How do generosity and service emerge in the lives of young people? How does such experience shape the long-term aspects of their lives?

- **The Science and Spirituality of Family Care-Giving**—Hundreds of millions of family caregivers across the globe are tending to loved ones with cognitive, physical, and other disabilities. How do spirituality and the ideal of love help to sustain such caregivers and those they care for?

- **Compassionate Love, the Analogical Imagination, and the Science of Creative Genius**—Creative geniuses have often attempted to express the ineffable reality of Unlimited Love in music, art, poetry, and prose (including theology). How can we better understand this human desire to express the ideal of pure love?

- **Spiritual Transformation: Love and the Fruits of the Spirit**—What is the relationship between positive spiritual transformation and Unlimited Love? How is the perception of such love captured in the experience of spiritual transformation?

- **Science, Religion, and the Metaphysics of Love**—People of all great religious traditions view Unlimited Love as a core aspect of the divine. Is there a scientific basis for this perennial assumption? Can it be clarified in the light of the sciences?

These questions were not all perfect, but at least an honest third of them were focused on God's Unlimited Love and our transformative experience of it. More than eight hundred people came from forty countries, representing a mix of premier scientists, renowned practitioners, and spiritual and theological thinkers from every tradition imaginable and then some. Sir John was unable to attend, but many people associated with the Templeton Foundation did. I mention these kinds of events simply to convey that Sir John's "humble approach in theology and science" shaped our activities and continues to do so to this day.

Summing Up the Provocative "Humble Approach" to Unlimited Love

It was a steep learning curve for me, and getting to the writing of *Unlimited Love as Ultimate Reality* has taken a number of years as I have gradually become more aware and appreciative of the full, bold dimensionality of Sir John's vision. My hope is that those who follow me will learn from my mistakes and therefore make faster progress.

In conclusion, Sir John asserted that we actually know almost nothing about God and spiritual realities like Unlimited Love. Unfortunately, though, we think we know more than we do, and this static arrogance blocks progress. He was skeptical about anyone who claimed to have a total grasp of the truth, for he felt this was egotistical and in contrast with his "humble approach":

> But the humble approach teaches that man can discover and comprehend only a few of the infinite aspects of God's nature, never enough to form a comprehensive theology. The

humble approach may be a science still in its infancy, but it seeks to develop a way of knowing God appropriate to His greatness and our littleness. The humble approach is a search which looks forward, not backward, and which expects to grow and learn from mistakes. (1981/1995, 35)

Furthermore, "The truly humble should be so open-minded that they welcome religious views from any place in the universe that is peopled with intelligent life" (1981/1995, 35). Like a tree, he analogizes, we can expect to know very little about the gardener; or like a clam we can expect to know very little about the nature of the sea (2001).

Sir John's embrace of a progressive, scientifically informed theological innovation had roots in the New Thought movement. As he wrote, "In recent centuries, hundreds of Protestant denominations have been born from new concepts and new revelations. Multitudes of cults and sects have arisen in other major religions, also. But how many of these sponsors research for more new ideas? The New Thought Movement, which includes The Unity school of Christianity and The Church of Religious Science, is a rare exception, one which strives for continuous innovation." (1981/1995, 60)

One finds in Sir John's writings a great respect for the traditions and scriptures of world religions, and yet he did not see them as sufficiently fluid and progressive:

The static viewpoint still hinders most religions. Neither the Koran, the Bible, nor the ancient scriptures of Asia say much about progress, and they say even less about research. Even now in the United States, the hotbed of research and progress, church bodies do little to help in the evolution of religious thought. They do almost nothing resembling the forward-looking research being done in the great scientific laboratories of corporations and universities. Within the different religious denominations of the church, the minor activities called research are essentially archeological, concerned with the excavations of ancient cities, the search for lost scriptures, or another translation of an ancient book. (1981/1995, 64)

Yet time and again he affirms these traditions and certainly affirmed the importance of regular worship (Templeton, ed., *Worldwide Worship*, 2000). His concern was that a purely "static viewpoint" would drive these traditions into obsolescence. (1981/1995, 65)

To overcome stagnation, the necessary ingredient is humility. He writes in *The Essential Worldwide Laws of Life* (2012), "Humility is the key to progress." He continues,

> Without it we may become too self-satisfied with past glories to launch boldly into the challenges ahead. Without humility we may not be wide-eyed and open-minded enough to discover new areas for research. If we are not as humble as children, we may be unable to admit mistakes, to seek advice, to try again. I use the word *humility* here to mean understanding that God infinitely exceeds anything anyone has ever said of Him, and that He is infinitely beyond human comprehension and understanding. As we realize this and become more humble, we reduce the stumbling blocks placed in our paths by our own egos. (2012, x)

"Humility," he notes, "opens the door to the realms of the spirit, and to research and progress in religion. Humility is the gateway to knowledge."

I return to Sir John's sense of urgency. He believed that in an age of ever advancing and, in some cases, ever more dangerous technologies and scientific prowess, the old tribalistic and insular forms of love will cause humankind more trouble than ever. Exclusive in-group love that arrogantly dehumanizes or even demonizes outsiders will result in unimaginable forms of harm and mass destruction. Sir John anticipated highly destructive forces of terrorism and surmised in the late 1990s that this difficulty might be the single greatest obstacle to world economic progress. Either we make progress toward an Unlimited Love for all

people without exception, or we will suffer increasingly severe consequences. Sir John remained always the optimist. He wanted progress in
Unlimited Love, and he was in earnest.

For Sir John, humility was not only the right approach to theological
progress. It was also the virtue that leads us to God inwardly:

> One of the major lessons to learn while on earth is that build-
> ing our heaven is up to us. Emanuel Swedenborg wrote that
> we will not be in heaven until heaven is in us. So how may
> we begin to build that heaven within? True humility can lead
> us into a prayerful attitude, and prayer can bring us in tune
> with the Infinite. There is a real mystical power in prayer, and
> it works. Through your prayer times and your attunement
> with God, you are increasing your own spiritual light. You are
> building a better expression of life in every way and you are
> attracting exactly what you are building—more light. There
> is a larger power you are touching. There is a larger life you
> are building. (2012, 134)

He notes that those who rely on "wisdom or beauty or skill or money
tend to shut God out," while those who are "humble and grateful" are
able to "open the door to a kind of heaven on earth here and now."
(2012, 135)

Humility for Sir John was never "self-deprecation" but rather "wisdom," "a flexible attitude," "knowing you are smart but not all-know-
ing," "the opposite of arrogance" (2012, 137). For Sir John a deep
spirituality underlies humility:

> Most great people are humble. Those among the most
> respected who have ever lived acknowledge that their greatness
> came, not from themselves, but from a higher power work-
> ing through them. The true meaning of humility is knowing
> that the personal self is a vehicle for a higher power. Jesus of
> Nazareth said, "I do not speak on my own authority, but the

Father who dwells in me does His works" (John 14:10). Other great spiritual leaders have recognized this; True genius has a deep sense of personal humility. (2012, 138–39)

He goes on to mention great Islamic and Jewish leaders, Sir Isaac Newton, Albert Einstein, and others who were known for humility and simplicity.

For Sir John, arrogance is an obstacle to all progress, as "the human ego also causes people to try to solve problems by human effort alone without seeking God's wisdom" (2012, 139). This ignores the power of God working with us: "Yet there is another part, a 'higher self,' that exists in each of us as a spark of the Divine. Unfortunately, most of the time this higher self remains hidden by the personal ego just described." (2012, 139)

So humility comprises the virtues we need to know God, just as humility itself is the virtue we need to make spiritual progress. Sir John identifies some great scientists having humility in both these senses, and he clearly associated visionary creativity with our willingness to serve as the conduits of a Higher Power.

Chapter Four

A Disciplined Rational Mind and the Power of Love Affirmations in Everyday Life

BEFORE WE MOVE AHEAD to Sir John's truly bold, visionary, "humility-in-theology" thinking about Unlimited Love as Ultimate Reality—the more essential and cosmic Sir John, the man who celebrated spiritual transformations and amazing grace experiences—we need to recognize that there was another side to him that focused in on everyday virtue and growth in love, and the benefits thereof with regard to happiness and success in a secular sense. That side of him reminded me of Emily Post, who believed that no American child could succeed without knowing table graces and a bit of etiquette and social virtue. This part of life was not what excited Sir John the most, but it meant a lot to him because he understood that society needs core virtue assets in order to function. Freedom without virtue and responsibility fails, and Sir John believed that freedom is God's gift—hence, his interest in "character education." Let us take the everyday Sir John with utmost seriousness, because everyday life and behavior meant much to him. However, keep in mind that by studying everyday human virtue, we are not studying the nature of God.

Sir John believed that human love, no matter how flawed, is still a limited expression of God's love in creation, a hint at something a billion times greater. He was a love spiritualist in that he intuited God's love behind all love. He often wrote simply of "helping," "kindness," "giving," and the like. "God is the source of all love," he wrote, "and if we open ourselves to receive his love, then we are able to radiate it to other people every day" (1987, 120). Thus, Sir John could question, "Do the qualities of unlimited love, joy, peace, compassion, and noble purpose that can grow in the garden of human hearts merely reflect the beauty and glory of the Creator?" (2003, 28)

Even the most natural human love—however complicated by impure motives, however ineffective and even counterproductive, however narrow in range and demeaning of outsiders, however dim and fleeting—is nevertheless something that points in the direction of a Creator who is the author of all love in the universe. Sir John wrote as follows:

> Could our relation to the Creator be like that of a sunbeam to the sun? Nothing can separate the sun from one of its rays. Made of the sun's substance, partaking of its nature, each sunbeam has a particular mission, a certain spot to caress and warm and light. Like the sunbeam, we, too, have our special spot to fill. We have our special work to do. Thus, we are part of the divine plan and necessary to the perfection and health of the whole. (2003, 13)

This idea of a "special spot to fill" highlights Sir John's notion that we are the right people in the right place with the right gifts at the right time to simply and genuinely contribute in love to the lives of those around us. In classical Protestant language he is referring to the "order of creation" in which each of us has a calling or God-given role in loving some constituency toward which we feel called, and doing so in a way that draws on our unique strengths. He wrote of the "ways of love" by which "Spirit can be channeled into any form of expression such as fine arts, crafts, healing, business, science, gardening, design, construction, caring for a family, and so much more" (2003, 118). All of this he felt was within a divine ordering of practical love in which no one is without some special gift that allows them to love others well.

Sir John did not contemplate the prosperity of the universe while neglecting the lives that surrounded him in ongoing relationships. Impartiality goes too far only when it delegitimizes legitimate proximities—the moral spheres in which most human beings spend much of their time: family and friendship. Sir John took wonderful care of his family and he had friends everywhere. The huge gathering of family members at annual Templeton Foundation Board meetings often included nearly one hundred relatives, including great-grandchildren, in-laws, cousins, and all. The amazing thing was that they all were so genuinely fond of him. They were a proof to me that Sir John abided in love with the nearest and dearest, which is often difficult because close up one sees all the details of imperfection. He never indulged his family members financially, preferring that they stand on their own and learn the virtues associated with success. His life was full of loyal friends who felt privileged to spend time with him. *He never said a bad thing about another person.*

People all over the world wanted to be associated with Sir John and his legacy because he was steadfast in love. He was simply the concerned good neighbor who hoped that a new paradigm of innovative and high-quality science focused on positive spiritual principles might further enliven our understanding and quicken our enthusiasm to abide in the ways and power of love. He claimed no ecstatic spiritual experience, no sudden conversion after years of sin, no moment of revelation when finally he was free from his past. Sir John was not a man who somehow discovered agape in mid- or late life in some dramatic moment of repentance and conversion; rather, he was nurtured during his rural Tennessee Presbyterian youth. Again, a farmer once remarked in retrospect, "That John Marks, he was born old," by which he meant that Sir John had a wisdom and maturity about him from his early years.

He wrote me a letter dated May 2, 2005, that seemed a slight respite from big bang and Ultimate Reality, as follows:

> Your wisdom will be welcomed by me and your other colleagues on the following five ideas:
>
> 1. Often people seek love to become happy. But really, happiness comes from giving unlimited love. The more love you can give the more you have left to give.

2. Unlimited love giving is over ten times more beneficial if focused more on ways to prevent rather than ways to alleviate suffering.

3. Giving materials often causes the receiver to remain childish, whereas opportunities to produce and to become self-reliant help the receiving to grow spiritually.

4. When a young child is able to listen, should a mother ask each morning, "How do you plan to use this new day to give more unlimited love and to whom and how?"

5. Life is always a vast mystery, but enlightenment can come from enthusiasm to give unlimited love.

God bless you.

So here we are reading the Sir John who was a cheerful giver concerned with overindulgence of children and the idea of a mother asking the right question to her young child.

Let's get back to a simpler discussion before we travel too far on the metaphysical journey or leap up too many electron rings. Sir John would not want heady reflection on Unlimited Love as Ultimate Reality to deflect the focus of our minds from the everyday encounters of a life well-lived in small acts of love, kindness, and giving. In fact he would wish to see any scientific breakthroughs in this area have significant impact at the cultural level where virtuous behavior in everyday life is formed. So we had better take Sir John's journey as a good neighbor now as a bit of an interlude and return to the biggest question immediately thereafter. Clearly, however, Sir John considered human love and human nature realistically, and he doubted if any human expressions of love in everyday life could long endure without some enchanting spiritual sense that our kindness goes with, rather than against, the grain of the universe and is consistent with the direction of Ultimate Reality.

Sir John did not believe that it was necessary to jump on an airplane and fly halfway around the world to manifest love, because we all have neighbors who are right here, in this place, in this moment. He wrote,

Start with whoever is around you—men, women, girls, boys, old people, young people, yourself. Express your love as a natural attitude and demeanor of goodwill, kindliness, support, caring, and benevolence, as well as a willingness to do what you can to be helpful and make things a little better for someone.

Giving love consciously—through thoughts, words, and deeds—can help you to become your force field of love. (2012, 160)

There is a resonance here with the Jewish idea of *Tikun Olam* (repairing the world), healing the world where you are, and having confidence that you have the right person in the right place to turn to and that the person who is right next to you can be loving. And start with the right affirmation!

VIRTUE AND SUCCESS IN LIFE

Sir John wanted people to be successful in life. He knew that no one stands much of a chance at success without common courtesy and a simple love of neighbor, kindness, and a smile—small acts done with a generous heart. I never detected anything snobbish or arrogant in Sir John. He was always concerned with all the wonderful everyday people he had grown up with as a youth, and he made a disciplined effort to focus his mind on treating every individual with equal kindness. He often attributed his success in life to the good people of Winchester, Tennessee, where he grew up. He "learned from his mother the content of his character could either lead to success or failure. His mother and Aunt Leila taught him how right thoughts can forge right actions." (1987, 16)

Sir John articulated the law of success by connecting love of others with happiness and flourishing:

Selfishness overlooks a key principle of success—helping others. Successful people meet the needs of others because it makes them feel good about themselves. Then, by subordinating any selfish motives to the greater motive of being of service, they successfully navigate their way through life. As with successful men and women throughout the world, our success is proportionate to the number of people we have helped to grow and prosper. (2012, 4)

This is a fabulous statement—success is directly proportionate to the number of people we help to grow and prosper. Sir John helped many people grow and many people prosper all over the world, and he was happy and successful as a result.

Being sincerely honest, diligent, respectful, reliable, and kind to the people around him meant a great deal to Sir John over the course of his life, and he considered this everyday love of neighbor to be the key to anyone's success in any domain of life and career. His hope was that every young person could realize that good things happen to sincerely good people, at least in general, and that the more they cultivate the virtues, the better off the people around them will be, as well as they themselves. He lamented the decline in the culture of virtue, civility, and learning that he knew wasted many young minds and lives.

The young Templeton as a Tennessee youth learned the simple practice of good neighborliness and hospitality. I asked Sir John's friend David B. Larson, MD (d. 2001), what he was like. Dave said to me, "Stephen, Sir John is the kind of person you would just love to have as a neighbor." I was a little surprised at this response, but I came to appreciate it as I came to know Sir John. In the sense captured in a Norman Rockwell painting of everyday goodness, he had great appreciation for the Golden Rule as a law of life that required no intensive spiritual experience but that had a universal validity (1987, 123). Sir John valued small acts of kindness greatly, and he manifested them. In such small acts he saw hints of an underlying divine love—and he recognized in each and every individual the image of God—so he tried to be a good neighbor to everyone he met. He had good role models as a youth in Winchester, and he was in turn a good role model to others.

Sir John was never caught up in theories and abstractions about the good and moral life. He had no need to bog down in meta-ethical theories, and he never wrote or spoke of them. Instead, his emphasis was on the virtues—and fortunately nowadays so many philosophy departments have acknowledged the severe limits of ethical theories and place renewed emphasis on the virtues in the modality of the new or neo-Aristotelianism. In general, a virtuous individual will probably do the right thing on the basis of experience with the concrete problems of everyday living, although good decisions do require moral luck; we never quite know if we did the right thing until we see how it all turns out in the proof of the pudding.

Sir John quotes Marcus Aurelius, "Your life becomes what you think about" (2012, 5). He encourages the "crowding-out technique" of filling the mind with good thoughts to displace bad ones. Here he even recommends an affirmation that he used: "You might affirm, 'I lovingly release you to the vast nothingness from whence you came.' Then just let them go" (2012, 6). This technique seems to have worked well enough for him that he could recommend it. For Sir John, thoughts tend to expand and grow. Like Charles Fillmore, Ernest Holmes, and Norman Vincent Peale, he suggested visualization techniques as a way of focusing the mind at deeper levels:

> In a short scenario, suppose you are faced with a complicated task, and your mind focuses on the word *failure*. Suddenly, an image might be evoked in which you fail at your task. This image could expand to the point where you may fail at other tasks and, possibly, to the point where people may ridicule you for your failure. Now, clear your mind, and visualize that you are faced with the same task, and decide to focus on the word *success*. Let positive images of accomplishing the task fill your mind. You see images of others appreciating your success, shaking your hand, smiling with admiration. This success image snowballs, and you can see yourself succeeding at other, more difficult, tasks. (2012, 7)

★ ★ ★

> Having such an image will encourage us to speak positively
> with others, and thus, our thoughts expand "not only within
> our own minds but expand through others as well." (2012, 7)

Sir John then quotes from Charles Fillmore again: "As you think,
so you are" (2012, 8). Everything in our lives—finances, relationships,
health, and happiness—are shaped by our thoughts. He cites Fillmore
further, "Thoughts held in mind produce after their kind" (2012, 12).

Sir John always emphasized loving thoughts as the best thoughts, in
combination with success and other primary positive thoughts:

> As you read this book, reflect on what your thoughts have
> created over a period of time. Defeatist thoughts, angry
> thoughts, dishonest thoughts, self-centered thoughts, and
> failure thoughts are destructive. Loving thoughts, honest
> thoughts, service thoughts, and success thoughts are creative.
> (2012, 14)

He recommends that we become self-aware of our patterns of thought
and correct them as needed:

> You can train your mind to nurture positive, loving, and
> unselfish thought patterns and, through them, develop a
> deeper, richer personality that may be the fulfillment and
> fruition of your greatest creative potential. (2012, 14)

We are responsible for what we think—and for our lives.

Sir John believed deeply in freedom and personal responsibility. This
belief stemmed from his emphasis on the subjectivity of the "inner"
world where "beliefs, thoughts, and feelings reside. Your happiness,
peace of mind, and enjoyment of work, friends, and loved ones often
depend more on this inner world than on the outer one" (2012, 17). We
are free to create our own reality through the self-control of the mind.

Our attitudes are ultimately our own. The "outer" world is filled with events over which we have little or no control:

> We have far more control over our inner world than our outer world. Not to say that changing our inner world is necessarily easy. We may have developed thinking and feeling patterns or belief systems that are deeply ingrained. Change may not always be easy, but it can be accomplished. Examining our beliefs and attitudes and observing our thoughts and feelings can be a useful place to begin. Change often starts to happen when we recognize false beliefs and make an effort to bring them in line with reality; when we recognize negative feelings and choose to give them no power over us. We have the power to create our own reality by choosing thoughts and beliefs that are positive and true. So, in truth, we create our own reality, our inner reality, the only reality in which we truly live. (2012, 18)

Sir John recognized that the "outer" world of circumstances can be difficult for us, but is it not the case that so many people in fact have been astonishingly successful in life because they have had to overcome obstacles? They have a certain advantage over people who have had everything handed to them and then fall apart in sloth, vice, and self-indulgence. As Sir John writes,

> "A person may be born in poverty, but by mastering and controlling his destiny through his thoughts, feelings, actions, and choices, he can rise above his limitations, and make the transition to a more spiritually evolved soul. Isn't this potential worth the effort to properly use the tremendous gift and power of your mind?" (2012, 19–20)

Sir John suggests a first step: "The first step is to pause right where we are, stop the chattering noise of our thoughts, and allow our thinking to adjust" (2012, 19). He is referring here to an emptying of the mind that contemplatives would typically conceptualize as "clearing" the mind through concentration on the breath, meditation, and prayer. Is it not true that our minds are too often filled with meaningless junk? The second step follows:

> A second step is to reaffirm your faith, lift your conscious-ness—and your thoughts—to a higher level of expression. Suppose you are experiencing some difficulty. Daily living may seem to be hard going at the moment, and you may feel uncertain about what you need to do next. Instead of beginning to panic and put pressure on yourself, pause for a moment and affirm, "Spirit goes before me, guiding and directing my efforts and my direction." Lifting your thoughts to a higher level can renew and restore the peace and serenity of your awareness—and your life! (2012, 19)

There is third and final step: "This is to go forward with confidence and courage, trusting the inner guidance that you receive and knowing that the way may indeed be made clear" (2012, 19).

I am reminded here of the work of Norman Vincent Peale, one of Sir John's friends from his New York days in the 1950s when *The Power of Positive Thinking* was a blockbuster, bringing New Thought themes into the mainstream Reformed Christian tradition of which Sir John was a part. Peale recommends a method to disciplining the mind: "The formula is (1) PRAYERIZE, (2) PICTURIZE, (3) ACTUALIZE" (1952, 42). Daily prayer is coupled here with "picturizing," by which Peale meant what we would now call *visualization*: "To assure some-thing worthwhile happening, first pray about it and test it according to God's will; then print a picture of it on your mind as happening, holding the picture firmly in consciousness" (1952, 42). The picture will tend to "actualize" by "invoking God's power upon it, and if, moreover, you give fully of yourself to its realization." (1952, 43)

Sir John is explicit in stating that his entire edifice of *The Essential Worldwide Laws of Life* is designed for happiness and success in freedom and responsibility: "There are certain good ideas, or laws of life, that we need to learn, practice, and cooperate with to lead a happy and successful life" (2012, 20). Moreover, "It is a law of life that whatever we give our attention to, and believe in, tends to become our experience." (2012, 20)

HOW TO? CONFIDENT AFFIRMATION AS THE PRELIMINARY STEP

We have to start a life of love with putting spiritual laws into words and speaking them or meditating on them with certainty and confidence in their truth. By speaking a spiritual law or principle as a "statement of being," by repeating an affirmation like "an attitude of gratitude brings blessings," we are self-training our minds. What we speak tends to take form. Commit affirmations from Sir John's *The Laws of Life* to memory and dwell on them with emotional depth and visualization. The calm affirmation of spiritual principles can work wonders in opening up the mind to an awareness of Divine Mind.

A saying in the popular culture is widely attributed to Sir John, but I cannot vouch for his having said it. Yet it does sound like something he would have recommended: "It's nice to be important but more important to be nice." I purchased a four-foot-long board with this affirmation engraved on it from a local gift shop in Port Jefferson, New York, in early 2012 and nailed it to the wall of my center office at Stony Brook University School of Medicine right over the top of the copy machine where people need something constructive to absorb into their minds. Such affirmations lie at the center of Buddhist, Hindu, and other meditational traditions. The affirmations are sometimes called "guided images" when they are spoken aloud and the listener imagines actualizing them in encounters. Perhaps this affirmation reached a few people here and there who could be a little nicer, and maybe if they were a little nicer, they could be a little happier. For Sir John, happiness is almost entirely a by-product of small acts of kindness.

Sir John suggested the use of affirmations, "especially when dealing with troublesome people" (2012, 166), and we all know some, and usually even a few! Sir John referred to these people as a blessing, as opportunities for spiritual growth. He refers to a businessman who pointed out to him the tremendous power of unconditional love. Before his meetings this businessman "filled his mind with the mental picture of people he was to meet with and blessed every one of them with an affirmation of love. Here is the affirmation: 'I am a radiating center of universal love, mighty to attract my good, and with the ability to radiate good to others, especially [the name of the clients].'" Sir John noted that affirmations generate a "powerful energy force field," and believed it important "to *feel* what you are saying; to feel the power of universal, unconditional love, pulsing through you and your words, with your whole heart, mind, soul, and strength" (2012, 166). In other words, affirmations with emotional depth have more impact on you and on God.

Affirmations were the basis of Sir John's efforts to impact the lives of young people. Sir John put tireless energy and substantial resources into the Laws of Life Essay Competition, focusing on students in grade schools, junior highs, and high schools across the United States. In 1987 he founded the first contest—in Franklin County, Tennessee—to encourage young people to write about the laws of life and how they have influenced their lives. These laws are simply affirmations, of which "It's nice to be important but more important to be nice" is one. He took affirmations with utmost seriousness because they are the way that we can train our minds for positive living. In twenty-five years, the contest has reached millions of youth, and organizations such as Junior Achievement China, Learning for Life, and the Georgia Rotary Districts Character Education Program operate large-scale Laws of Life contests. Junior Achievement China's goal in 2011 was to reach one million students through these contests.

Unlike the Kantian who believes that our minds need to be "purified" of content to apply *ex nihilo* the power of logic to arrive at something like the Golden Rule, Sir John believed that in fact every mind is filled with something, and that our primary discipline is to "crowd out" the negative thoughts with positive ones through affirmation techniques. Think of the mind like a computer that has to be running on some software program or another. The question is only what we choose to upload in our individual freedom but as creatures of culture. In my view, this is a truer sense of the mind than we see in Kant's "pure reason," and in the tabula rasa (blank slate) model of mind. We are all constantly under the influence of culture, relationships, and our own emotional states. We are free but not all that autonomous. Our autonomy comes when we decide as individual agents to concentrate our minds on loving and creative affirmations and to crowd out the various alternatives, which is a hard thing to practice consistently because our minds tend to slip down into places where we should not allow them to go.

Everything begins with an idea, with a thought. Nothing begins without one. Every piece of architecture, every painting, every book or piece of music, every meal prepared or word spoken begins with a thought. Indeed, Sir John asserted that each of our minds is a small drop of the ocean of Divine Mind, and that Divine Mind is the one that began the universe with thought and Word. "Let there be light," and there was light.

Sir John was anything but bored with the details of everyday encounters. He once asked me if I thought that researchers who study love and Unlimited Love need to make a diligent effort to be role models for others. I responded that when it comes to a topic this big, this contrarian, and this challenging, it is especially important to sincerely try to live a life of love in every encounter, even ones with people you may only encounter one time in life. Sir John vehemently agreed. However cosmic Unlimited Love may be, however much it is an energy that may be a universal, unseen force like gravity, our lives in it must begin with minds focused on loving encounters in our families, friendships, neighborhoods and communities, schools, workplaces, and in any other domains of everyday activity. God and providence have placed people in our immediate local lives.

THE EVERYDAY DISCIPLINE OF THE MIND

Sir John stressed the power of the disciplined mind as a source of behavioral guidance. He believed that God gave us our minds to use them well by freely self-inculcating the right kinds of affirmation, such as the

Golden Rule, or a saying like "An attitude of gratitude brings blessings." Mind Power was very important to him, and he did not like to see people wasting their minds. No, the unaided human mind is not capable of the spiritual transformation to God's Unlimited Love, which is a much more profound matter. Yet we cannot afford to ignore Sir John's sense that we can at least achieve more consistent, wise, and fruitful love through the right training of our minds. With this in mind, he wrote *The Laws of Life* and edited it several times over two decades.

The "love of neighbor" was at a certain level for Sir John a command, a principle, or a law of life. This is not the love of neighbor that arises from a spiritual transformation or from some experience of Unlimited Love (a.k.a. "God"). This more mundane love of neighbor might involve emotional warmth, a rational argument for equality and equal-regard, an innate helping impulse, or maybe the right role model. But it does not make a quantum leap upward into that Unlimited Love that is God and that is millions of times greater than humans are capable of, even when they focus their minds on "Do no harm" and "Do unto others." The mind must be trained to treat all people with beneficence and respect. Yet Sir John still felt strongly that the human mind, at least in its spiritual essence, is connected to the oneness of Divine Mind, and an awareness of this connection is part of mental discipline.

God has given us our minds. Because each mind is a gift, we must each value it and train it in these directions. Thus, we must train our minds to love everyone we encounter as in the image of God, and as significant and worthy of our concern.

Sir John often quoted this affirmation: "When you rule your mind, you rule your world":

> If you desire to understand the reason behind the statement, "When you rule your mind, you rule your world," take a look at what some religious and spiritual philosophers call Infinite Mind and the Law of Mind Action. Some say there is only one Mind, sometimes called Spirit, or God Mind. This Mind is the life, intelligence, power and creativity that suffuses the

entire universe. Yet they say the Law of Mind Action holds that we are individual and yet remain a part of the whole. (2012, 3)

Here he is stating that while our individual minds are on the one hand like drops in the water of Infinite Mind in which each of us participate, each of us are also individuals with our own identity and agency in the world and in eternity. In other words, we are connected in various ways with the Infinite Mind that gives us part of that Mind as our own, yet we are also discrete entities responsible for using our minds responsibly and to the fullest. Yes, "We have free will" (2012, 3). But because our individual minds do merge with Infinite Mind, the power of our minds is so much greater than we might imagine. Note that, following Fillmore, Sir John equates Mind and Spirit in the above passage.

To unlock the tremendous power of the mind that is to some degree of one substance with Infinite Mind, we must use positive thinking and discipline ourselves to do so. No doubt Sir John believed in the power of positive thinking. In fact, positive thinking is what we should be doing with our minds, in contrast to negative thinking. Positive thinking is the key to unleashing the power of our minds, which is astonishingly great—because in some part we are participants in Infinite Mind. Sir John writes,

A positive attitude toward life can be difficult for some people to adopt, for it may seem unrealistic. These skeptics may find it hard to believe that positive thinkers can accomplish almost anything they set their minds to. But, with a positive attitude, your chance for success in any situation is greater if you look for workable solutions rather than allowing negative thinking to limit your decision making. (2012, 4)

In so many ways, love is for Sir John an attitude of spirit or mind, rather than anything explicable in terms of emotion.

"Are You in Control of Your Mind?"

In December 1962 the message on the Templeton family Christmas card was as follows: "On the 1962nd birthday of Christ, we invite you, our friends, to share with us this little simile: ARE YOU IN CONTROL OF YOUR MIND?" While a tad unusual for a Christmas card greeting, it is filled with wisdom. The expression "thought control" has an ominous ring to it because it is associated with control from outside the self, and as therefore contrary to freedom. But Sir John's approach is the total opposite, for he focused on self-control from within as a matter of mental and emotional discipline. James Ellison concluded, "His theory of positive thought control is actually a deep form of self-control." (1987, 111)

Each of us, as free individuals, can make a 100 percent effort to fill our minds with thoughts that strengthen and expand love. Despite all the negative influences and cultural assumptions that can possess our minds, we alone are ultimately responsible for filling our own minds with the thoughts that can make our lives better and more loving, and free of resentment or other destructive emotions. As Sir John noted,

> Each word or action begins as a thought. We can say loving words and do loving deeds only if our minds and hearts are full of love. Is it good to stand watch over each thought that sprouts in the mind? If it is a loving thought, let us cultivate it. If it is not, can we crowd it out by filling our mind with loving thoughts? Try this experiment: Think of any person you envy or resent. Then free the mind of that poison by seeing that person as a child of god, and pray for that person's welfare as an experiment in self-discipline. The divine dwells in various ways in every human being, so it is easy to find good qualities in any person if we lovingly try. As Jesus said, we should pray for those who irritate us. (2000a, 142)

This concept of crowding out negative thoughts by cultivating positive ones is to be found again and again in Sir John's writing. The mind is always filled with something. It is never empty. The key concern is what we each allow into our minds, for this determines our attitudes and actions. In a vivid statement, Sir John also wrote this especially important passage:

> Our mind is very much like a garden, fertile with good soil, water, sunlight and drainage. We, as the gardeners of our minds, can cultivate whatever thoughts we choose. Should we nourish good thoughts, weed out the bad ones and ensure that evil thinking does not overshadow and block out the radiance of the good? By years of careful thought control, your mind can become a garden of indestructible beauty. "As a man thinketh so he is." (2000a, 142)

Sir John was astonished, inspired, and awed by the human mind. He viewed it as the great miracle and mystery of creation and of the universe, and possibly the result of a God for whom the presence of such minds makes possible a relationship of love. Sir John asked, "Are we privileged to be so wonderfully made to have minds arising out of bodies that allow us to move and sense and participate in the rich reality that surrounds us?" (2000a, 25). Sir John could hardly believe the amazing complexity and power of the brain:

> It is encouraging to realize that the most complicated object that we are now aware of in our universe is the human brain with its as yet mysterious property of inner emergent consciousness. Consciousness combined with intelligence and an extensive capacity for memory makes us "us." It provides us with the basis for being free agents with the ability to communicate, to know and interact with other similar persons and therefore to live in communities and participate in shared knowledge and activities, to be creative individually and to love, to dream, to hope and also to have deep and abiding spiritual desires and potential. We are not yet aware of how

this aspect of reality arises within what biologically is simply
a special form of flesh. (2000a, 24)

That the mind could have a spiritual thirst was nothing short of astonishing to Sir John.

At some level, the life of love is a decision, which is to say that we make
up our minds to do it because it is better than the alternatives. Sir John
approached mind through the use of affirmations. This tradition, in the
broadest sense, harkens back to Marcus Aurelius, "Your life becomes
what you think," and carries forward to Charles Fillmore's statement,
"Thoughts held in mind produce after their kind." Sir John asserted,
"We must train ourselves to genuinely and deeply love every person;
we will then, in return, be people who are magnets for love" (1987, 39).
He recommended the following practice: "Visualize unlimited love in
expression and develop greater attunement with the quality you are
picturing in your mind. The experience of the presence of the Creator
is within us and available to each of us at all times." (2003, 89)

Sir John believed deeply in freedom and responsibility, and clearly
each of us is responsible for our mind, and practice we must:

> It cannot be repeated often enough that we can learn to radi-
> ate love, but first we must practice using loving words and lov-
> ing thoughts. If we keep our minds filled with good thoughts
> of love, giving and thanksgiving, they may spill over into our
> words and deeds. If we are not very careful to weed out all
> evil thoughts, such as envy or hate or selfishness, they, too,
> may overflow into our words and deeds. To produce beauti-
> ful music requires long practice, and so does the production
> of a beautiful mind. With practice, both come more easily.
> (2000a, 147)

The self-control of thoughts underlies everything that we as humans
create in an otherwise chaotic world:

Obviously, to build a house we begin with thoughts, then words, then deeds. Most good objects produced by humanity are created by this process, which starts with thoughts and then develops into words and deeds. Nations are formed in this way, and so are sciences, as are all organizations and institutions of human society. Even more awesome is the fact that thoughts can build not only outwardly but inwardly. By thoughts we create not only our possessions but also our personalities and our souls. By the long practice of controlling our thoughts, can each of us make ourselves the kind of person that ultimately we want to be? (2000a, 146)

The thoughts that we cultivate through concentrated affirmation over time will shape our present and future lives as surely as any great work of sculpture begins with a mental image before a chisel can be taken to a block of marble. Our thoughts are carried forward into words and actions, so what we allow to dominate our minds manifests in our lives. If we focus our minds on unconditional and universal love, we have the capacity to extend and intensify our love so that it might more fully resonate with the equal-regarding and exceptionless love of the Creator.

Sir John did not need to quote anything more than his beloved passage, *"For as he thinketh in his heart, so is he"* (Proverbs 23:7). He saw value in many methods of focusing the mind on the laws of life, including rituals, prayers, meditations, music, art, verse, or forms of worship. He collected such methods from around the world in an impressive edited book, *Worldwide Worship: Prayers, Songs, and Poetry* (2000). Indeed, world religions at their best are designed to prime our minds and hearts for living according to the law of love. Of course, he knew that religions can sometimes bring out the worst as well as the best in people. He wanted to see the best.

Sir John did freely pursue his visions for improving the world. He knew that many philanthropists performed marvelous work in contributing to needy organizations that tend to the weak, infirm, and vulnerable. He too was generous in helping others. But his bright and creative vision was to help all people, without exception, by slowly bringing the world to greater knowledge and practice of the laws of life, of which love is the most overarching. Sir John knew that with these principles of living within each of our minds, and experimented with in our daily lives, we could flourish at every level—interpersonally, emotionally, physically, economically, and spiritually. He felt that engaging the law of love is crucial to human well-being and progress, and that the expansion of these principles into lives and culture is absolutely imperative for the future of civilization.

The principle of *The Essential Worldwide Laws of Life* is that our minds and thoughts are tremendously powerful in shaping the reality around us. This principle and spiritual discipline made its way from New Thought into Protestant Christianity through Norman Vincent Peale's *The Power of Positive Thinking* (1952). Sir John knew Peale personally for many years. It made its way into the business world through Napoleon Hill's 1937 classic *Think and Grow Rich* (1937/2005). It resonates with the mainstream "rational emotive therapy" movement of Albert Ellis. Sir John invited youth into this way of life. Indeed, in 1997 he handed me an autographed copy of *Worldwide Laws of Life* (1997) to give as a present to my then fourteen-year-old daughter, Emma.

Sir John—and New Thought generally—have always stressed the centrality of agape love. New Thought pioneer James Allen focused on the theme of agape love as the chief source of happiness. In his 1903 classic *As a Man Thinketh*, Allen articulated a line of thinking central to Sir John as follows: "The heart that has reached utter self-forgetfulness in its love for others has not only become possessed of the highest happiness but has entered immortality, for it has realized the Divine" (1901/2008, 122). Allen continues, "Lose yourself in the welfare of others; forget yourself in all that you do; this is the secret of abounding happiness" (1901/2008, 125). Certainly, focusing the mind on the ways in which we can love our neighbor (through visualization techniques, prayer, worship, reading inspiring stories, rituals, and the like) is a crucial element in preparing the human agent for the spiritual experience of Unlimited Love. I was invited to give an annual address in 2006 at the Unity Cathedral in downtown Kansas City, Missouri, standing under the stained-glass window designed by Charles Fillmore. It is a simple window—a blue background with the word "love" in red capital letters. On the bottom floor is a bookstore dedicated to his works, which are most easily available in *The Essential Charles Fillmore: Collected Writings of a Missouri Mystic* (1999), considered the definitive edition of his writings.

The power of mind can expand the range of human love and is perhaps powerful enough to achieve degrees of equal regard with another and a sense of a shared humanity. Our thoughts themselves may be "things" with influence in a universe where energy and matter are one. Our thoughts certainly impact our words and behaviors. Anything that we can achieve through the power of our minds to extend love to the level of equal regard is welcome. So we must love in thought, word, and deed, knowing that thoughts are things. Sir John understood that the mind is always being shaped and programmed by some set of ideas, and there is really no such thing as an "empty" mind. The only serious question is what we fill our minds with, and that makes all the difference.

THE LAWS OF LIFE SAVING YOUNG MINDS IMPERILED BY DESTRUCTIVE CULTURES

Sir John encouraged "generally accentuating the positive ideas and attitudes and avoiding the negative." He was concerned about the negative thinking that permeates our contemporary cultures through horrific lyrics and debased language. His desire was to liberate young people from this negative cultural vortex. Sir John took special interest in young people, and he wanted them to be able to discipline their minds:

> Religions generally agree that "as a person thinks, so is that person." If this is an authentic and demonstrable spiritual law and if it can be taught convincingly, especially to young people, it might be the basis for new generations much more

disciplined in the control and management of their minds and
lives than current generations. (2000a, 159)

Perhaps the best way to capture this aspect of Sir John's thought is
through his *The Essential Worldwide Laws of Life*. This book is taken
from Sir John's *Worldwide Laws of Life: 200 Eternal Spiritual Principles*
(1997). *The Essential Worldwide Laws of Life* is all Sir John, but is now
organized thematically into chapters and reduced a bit in length by the
elimination of inessential embellishments here and there. Though Sir
John passed away in 2008 he was always looking for ways to present his
laws of life more effectively, and I feel certain that he would have been
ecstatic with this new rendition.

There are twenty-one thematic chapters of *The Essential Worldwide
Laws of Life*. Chapter 1 is titled "Turning Thought into Action." The
first law included there is in the form of a quote from Bill Provost,
"When you rule your mind, you rule your world." Here we find the great
idea that we are free to become what we think, that positive thinkers
can accomplish almost anything they set their minds to, but that "Selfishness overlooks a key principle of success—helping others." Indeed,
success is proportionate to the number of people we have helped. Often
each chapter includes up to two dozen laws. Three others in this chapter are "As you think, so you are" (Fillmore), "Thoughts held in mind
produce after their kind" (Fillmore), and "Change your mind to change
your life" (John Marks Templeton). He begins, "What does your mind
have to do with your life? Everything." Each law is described in easy
language that any junior high or high school student can read, but that
any adult reader can gain from equally. Wonderfully engaging vignettes
and stories are used to get the point across. Sir John was a great believer
in the power of story and narrative, and he encouraged me personally
to tell good stories when I write because people love stories.

The Essential Worldwide Laws of Life is designed to present opportunities to focus the mind. The thoughts of so many people, young and
old, are scattered and pulled into disarray by the drawing power of the

material world around us. Our glitzy world attracts our minds through the senses. But Sir John recommended a focus on the "inner world" of the Mind, and he charged each of us to choose wisely the quality of our own thoughts. Such good "gardening of the mind" allows for the mind to develop the concentration and power needed to optimize its creative abilities. In this process of diligent stewardship of the mind, Sir John felt, we can realize or become aware of a deeper connection with Divine Mind and better recognize the value of our own being as made in the *imago Dei*, the image of God. We are more likely, in the calm of a controlled mind, to sense the unconditional, unwavering, and no-strings-attached love that God has for each of us. When we control our minds, we are able to realize the inner quality of pure love that is part of our original nature, no matter how much life's difficulties have repressed this. In a self-controlled mind, focusing on noble affirmations and values, we are at last inwardly free, and the mind can extend love more purposefully and widely into the external world. With a controlled mind we are more likely to find the secure base of Unlimited Love that is available to us; we can find wholeness or inner completeness, and we can become a lighthouse to the world. But to make these internal discoveries we need to exert our supreme effort to center our minds on truth and love.

Given how much Sir John appreciated the Love-Gift of the mind, few things troubled him more than wasting it. Thus, disciplining the mind and filling it with positive and creative ideas like a gardener planting fruitful seeds was for Sir John a mark not only of gratitude and good stewardship but the only way to live a flourishing life. He saw many young people who were not using their minds well, and thus he tried to teach them his laws of life—thought affirmations intended to prime the mind toward fuller development. We should each have affirmations of love not just for our nearest and dearest but for the neediest and for everyone on the earth now and in the future.

Sir John believed that "God loved us so much He gave us our mind." In his *Worldwide Essential Laws of Life* (2012) he asserts these principles:

- Our thoughts are carried forward into manifestation in our daily living.
- The thoughts that will be most helpful to us are centered around such things as visualizing the future in clear terms, helping others, forgiveness, humility, gratitude, noble purpose, courage, joy, unselfish love, self-control, forgiveness, perseverance, and humility.
- We are each ultimately responsible as free individuals for what we allow into our minds and hence for creating our lives.

Of course Sir John appreciated that our ability to love is influenced by role models who exemplify this virtue in everyday life. He was fortunate to have a number of such exemplars growing up, and most of us do. We hopefully retain the details of their interactions from tone of voice and facial expression to some small act of kindness that sticks in our memories. But Sir John also emphasized, and even more so, how we can extend love and make it more unconditional by freely controlling the thoughts that we allow to occupy our minds.

Sir John truly believed that if each of us can rule our minds and sustain noble thoughts, we can live better, more loving lives. But even the most focused mind cannot arrive at Unlimited Love, or this requires the deepest connectivity with Divine Mind in its mystical Unlimited Love.

Chapter 5

The Spiritual Mind That Can Experience an Absorbing Field of Pure Unlimited Love That Far Transcends Ordinary Human Love and the Rational Mind

As the previous chapter makes clear, Sir John thought that the more we focus our minds on loving others—the more we approximate an equal-regarding unconditional love for all—the more likely we are to reach a point of resonance with the energy of God's Unlimited Love. Like a windmill, at this higher level we can catch the wind and enjoy a whole new energetic gust of Unlimited Love as an energy field, a field of love infinitely greater than frail human love alone. Sir John felt that as we do our utmost to abide in human love, we draw the Divine Mind and the power of Unlimited Mind closer to us.

Sir John understood the human creature as a threefold being, made up of spiritual mind, rational mind, and body. The *spiritual mind* is that part of our mind that is eternal. It is a drop of God's Mind within us, the same substance as Divine Mind. This is that inner dimension of spirit that the great sages refer to as "God within." Those who experience it speak of a loss of any sense of time or space, of a sense of oneness with all that is, of a sublime warmth and light, and of the deepest tranquility. This part of us is the point of contact with God's exponentially greater love and creativity—the part of us that is empty without the cultivation of God's presence and that is the root cause of all human questing to make "spiritual progress," to know and feel the living reality of God, and to pray. It is this spiritual mind that can only find "rest" in God, and that can receive creative inspirations and intuitions from God. This is the aspect of mind that experiences dramatic spiritual transformations. It has been called "Cosmic Consciousness" in its merging with Ultimate Reality in a manner accompanied by intense bliss and ecstasy. This spirit mind experiences a connectedness with all of reality at the level of its underlying primordial matrix. This is often experienced as a kind of invasion of moral consciousness, as being touched by Ultimate Reality as Unlimited Love. This involves that part of the mind that is God within us. This is the aspect of our minds that is non-local and nonreducible to matter. William James, Wilder Penfield, John Eccles, Maurice Bucke, Walter Stace, Pitirim Sorokin, and so many others have affirmed such an aspect of our minds, as have all the mystics, and many have said that increasing awareness of this will be the basis of the next great stage in spiritual progress. Sir John once supported a project in spiritual transformations, and he understood how profound the transformations can be that involve an opening of the spiritual mind.

Then there is the *rational* mind of conscious thinking and intellectual discipline that we discussed in the previous chapter. How important it is that we develop our rational minds and not waste them. With our rational minds we can formulate a vision for the future and implement it, we can articulate cognitive affirmations and abide in them, and we can achieve great things. Sir John appreciated nothing more than rigorous intellect, scientific method, and conceptual clarity.

It is in the spiritual mind that the human creature experiences the gift of a small grain of the Divine Mind. As he wrote, "For the sake of clarity and simplicity, some analysts think of each person as composed of four basic components: god, soul, mind and body" (2000a, 142). What did Sir John mean by this? He endorsed the idea "*that each soul is immortal and can be educated*" (2000a, 142). Here he has in mind "the soul" or "spiritual" mind in its eternity and of the same substance as the Godhead. The rational mind he describes as a "strategic link between soul and body" that is "emergent within the flesh of our bodies" (2000a, 143). "*Mind is complex and miraculous but temporary*" (2000a, 143). He stated the following:

Is it far better to have a powerful and/or beautiful mind than a powerful and/or beautiful body, and far better still to have a beauty of spirit in ways that join with the eternal divine love? (2000a, 143)

Sir John most definitely recognized a point of connection between God and the eternal human soul, and in this sense, he believed that we are all part of God. This does not require some ecstatic experience but rather is a matter of insight. This connectivity is always there, almost like an object that we have misplaced.

Sir John wrote, "This is the wonder and the mystery of it, that when we love God we get an enormous increase in the quantity of love flowing through us to others" (2000a, 159). He always kept the two halves of the double love commandment (Matthew 22:37–40) together, which means that he did not want to pretend that the ethical half can ever really be implemented without the prior spiritual half:

> Thou shalt love the Lord thy God with all thy heart, and with all thy soul, and with all thy mind. This is the first and great commandment. And the second is like unto it, Thou shalt love thy neighbor as thyself. On these two commandments hang all the law and the prophets. (2000a, 160)

And immediately Sir John offers an extensive interpretation:

> This can be researched as a basic law of the spirit. A person who applies this law often finds her life revolutionized. Opening our hearts allows God's love to flow through us like a mighty river. If we love as God loves us, we learn to love every person without exception. The happiest people on earth seem to be those who give wholeheartedly always. (2000a, 161)

Sir John sees humility as key in gaining this connection or in becoming more aware of it. "*If we get rid of our ego-centeredness, we can become clear channels for god's love and wisdom to flow through us, just as sunlight pours through an open window."* (2000a, 34).

At a deep level, Sir John affirmed that we can be channels of the Creator's love, and not just reflections of it. Thus,

> God is the source of love. Love cannot flow in unless it also flows out.
>
> The Spirit of God is like a stream of water and His disciples are like many beautiful fountains fed by this river of waters. Each one of us is such a fountain, and it is our task to keep the channel open so that God's Spirit can flow through us and others can see His glory. Without God, we are not likely to bring forth any good. If we think too much of the visible world or trust in our own ability, we become a clogged fountain. We will never learn to radiate love as long as we love ourselves, for if we are characterized by self-concern, we will radiate self-concern. (1981/1995, 110)

Here one sees how meaningful the idea of being an open channel to Divine Spirit and Mind was to him.

There is little doubt that Sir John assumed the existence of a higher divine love that transcends all human capacities: "Unlimited love was called agape by the ancient Greeks to distinguish the divine love from earthly emotions" (2000b, 3). Sir John saw a small spark within each of us that is waiting to ignite us to new levels of love:

> *The Bible tells us we are "made in the image and likeness of God."' What does this mean? How would we describe our perception of the image and likeness of the Creator? Could this mean the divine presence resides as the deepest and most intimate spiritual reality within each human personality? Could this inner-spirit spark radiate the Creator's unlimited love and creativity directly through each one of us?* (2003, 118, italics added.)

In another passage he wrote,

> *Many have taught that those who worship the creative Source are empowered by the Spirit and spiritual exploration and that growth can be a means of being lifted to new heights of joy, love and philanthropic achievement. The divine Spirit moves into our lives and makes us over from within so that all things are seen in a new light. Unlimited love often becomes a spontaneous expression of a spirit-filled soul.* (2003, 16)

These experiences bring joy, something deeper than happiness: "Like the sun, its joy is manifest in the shining forth of its nature" (2000b, 6). "Have you ever noticed that some people seem to be happy no matter what may be taking place in their lives? There is buoyancy to their spirit and a sparkle to their personalities. A kind of glowing energy field seems to radiate from their faces, their words, and even the way they walk! What is the source of their inner radiance?" (2000b, 41). Here Sir John has moved far beyond the rational mind and its limitations. *However much reason primed with the right affirmations can achieve for the life of love, it does not arrive at the Unlimited Love that invades our spiritual minds.*

I cannot be more recise than this and remain faithful to Sir John's writings, rather than going beyond them. He did understand all mind and consciousness as nonmaterial, although he seems perhaps to have linked the rational mind more closely with the body and brain, and asserted its relative finitude.

The Mind of God is Infinite. The spiritual mind of each individual is some part of this Infinite Divine Mind. This is where the greatest expressions of creativity come among those who feel more or less astonished and invaded by some insight discerned in a dream or in a moment of bright light. Consider the creativity of a Michael Faraday, who literally had visions of electromagnetic radiation and made great advances without knowledge of mathematics. The rational mind can take these divine intuitions and work with them. But our spiritual minds can tap into the reservoir of the Infinite Mind in ways that reason alone certainly does not. Each of us can connect with the Mind of God that is at the center of our being. *The spiritual mind is that part of us that experiences oneness with the Ground of Being, with Ultimate Reality, with what mystics east and west, north and south, have been experiencing directly and without human intermediary interference.* Here we are quantum leaps higher than the rational mind.

So back to the perennial philosophy: we, Atman, are of the same stuff as God, Brahman. Divine Mind is in us. Again, let us acknowledge the perennial philosophy:

> The divine Ground of all existence is a spiritual Absolute, ineffable in terms of discursive thought, but (in certain circumstances) susceptible of being directly experienced and realized by the human being. This Absolute is the God-without-form of Hindu and Christian mystical phraseology. The last end of man, the ultimate reason for human existence, is unitive knowledge of the divine Ground—the knowledge that can come only to those who are prepared to "die to self" and so make room, as it were, for God. (Huxley 1994/2009, 21)

Sir John clearly understood that an aspect of each of us is part of a greater Divine Mind capable of creativity, insight, and love that far surpasses the capacity of the human mind unplugged:

> The good news is that, even though the light of creativity and knowledge may be suppressed and hidden under innumerable bowls of belief we place over it, it never loses its divinity or its immense and indefatigable energy. It will try to work through the subconscious facilities instead, and we may find ourselves having interesting dreams that make a deep impression on us. (2012, 79)

He writes of our being "open to the infinite mind of God in universal form. Who knows what realms of knowledge and learning abide therein, waiting for us to be receptive" (2012, 78–79). He recommends that we listen to our "inner promptings" (2012, 79). Here Sir John is far beyond the rational mind with its discipline of affirmations, although he clearly thought that a well-disciplined rational mind was vitally important and in many respects elevating to the point of making the spiritual mind come into awareness.

ONENESS: UNLIMITED LOVE AS AN ASPECT OF SPIRITUAL MIND

On the one hand, Sir John understood naturally evolved human love is an indirect participation in God's love, but leaving the discussion there does him a profound disservice. He was not asserting any deep convergence between natural human love and Unlimited Love. Human love is often deeply distorted by wrong thinking, selfishness, and various corrupting influences. Sir John was not naïve about human nature; he read the papers. Human love is always to some degree—and often to a great degree—impure. Hence his book title is *Pure Unlimited Love*, by which he meant that each of us can only know truly pure love by directly participating in the fountainhead of Unlimited Love, the Ultimate Reality. This love is not tinged by the muddy waters of the baser aspects of human thinking but is rather fresh from the original Divine Mind. It is as though all love is part of a stream flowing from a pure source, but as it flows it becomes muddied and impure, which is how we know love in human interactions. *Pure Unlimited Love* refers to the original source, high up in the mountains, before any corruption by erroneous thought and selfishness interferes with its clarity and beauty. The original pure love is also pure in the sense that it is undiscriminating as it simply pours itself out unconditionally. Love in all our human relationships is on the one hand a reflection of the divine, but it is also diluted and impure. It is muddy and downstream. So the spiritual mind can be purified, enlivened, and quickened when the mighty currents of Pure Unlimited Love stream forth from the fountainhead of Ultimate Reality.

We must acknowledge Sir John's interest in the perennial idea that all minds are connected and participate in a Divine Mind. This idea of an aspect of mind in each of us that is beyond time and space and one with the mind of God is definitive of the perennial philosophy. Is it such a strange idea, or have we just been too frightened to take it seriously because a materialist view of the universe insists that every aspect of the human mind is merely the localized product of the brain? Are our lives all interconnected because at some level our minds all connect together in the One Mind?

Consider those intuitive hunches when we just "know" something, as if we are tapping into a higher level of knowledge. Consider the synchronistic "coincidental" encounters that seem to be directly in answer to a prayer. Consider people's surprising premonitions and the dreams that turn out to be accurate. Carl Jung could not deny the pervasiveness of these sorts of phenomena, and so he posited the existence of Divine Mind in the form of a Collective Unconsciousness.

I told Sir John my personal experience with nonlocal mind came when I was a freshman at Reed College. One night I made the mistake of accepting a ride on a Harley-Davidson from a visitor to campus whom I had never met. It turned out he was absolutely out of his mind. He accelerated to 120 miles per hour in about two minutes, ran all red lights, and headed south on the Pacific Coast Highway for an hour before doing a screeching U-turn. He returned at this blazing speed and dropped me off near campus. This was a typical rainy Oregon night, and all I honestly thought was that I was a dead man. I stumbled into Akerman Dormitory, frozen and frightened to near death. It was 11 p.m. As I opened the door to the dorm and entered the common room, the telephone mounted on the wall rang. It was my mother, who had woken up at 2 a.m. in New York with a terrible sense that I was in the gravest danger imaginable. "Stevie," she asked, "Are you alive? I had a terrible feeling that you were in great danger!" I responded, "Mom, I think I came closer to death than ever in my life, but I am okay. How did you know?" She answered, "Moms know." I thought to myself that my mother had a connection with me because, despite all, she loved me so much. Maybe deep love somehow allows the mind to transcend the limitations of the body and reach out across space. I discussed this story with Sir John in one of our personal meetings, and he was impressed and wondered aloud if love is the energy that somehow allows such events to occur, tapping into some aspect of the mind that is active in Unlimited Love.

Maybe we just do not realize how nonlocal the mind is, or maybe we are just afraid of being scoffed at. Psychiatrist Carl Jung parted from Sigmund Freud because Jung disagreed with the latter's purely local view of mind. Jung's "collective unconscious" is something in which he felt we all participate, and hence his "archetypes" and "perennial forms," which are not all that different than Plato's realm of ideas. We seem to repress serious discussion of nonlocal mind, but a contrarian like Sir John could at least raise these big questions without fear of the materialist naysayers. In the world of philosophers of mind, there are many who assert some form of dualism along the lines of Thomas Nagel. If there is some *nous* or original mind as Nagel posits, then we in fact are all in some manner participants in it, and thus our connectivity in Ultimate Reality is not to be dismissed as entirely implausible.

Sir John did not believe that our minds are walled off. On the side of the nonlocal mind, which is the same as the eternal soul, were psychologists and mystics including Jung, Rilke, and Meister Eckhart, as well as physicists like Schrödinger and Henry Margenau, not to mention David Bohm's argument for a "holographic universe." Sir John alludes to all of these figures in his writings. So the Ground of Being and the Ground of each of our minds may be one. No, Sir John did not claim that modern physics or neuroscience proved it, but he clearly moved away from the distant God "above" to the immanent God of the perennial philosophy, and he honored the common human experience of nonlocal mind. He very much surmised that we are all connected and that we can experience connectivity with the Divine Mind, although we are not all aware of this:

> Perhaps in the astonishing capacity of our minds and in the
> spiritual hunger that so many people can describe vividly, we
> have some form of a link or a bridge whereby some human

minds may resonate with the mind of god in various subtle ways. (2000a, 25)

It seems that Sir John had an appreciation for the connection between God and the human being in our minds, one especially likely to manifest in those who love God and, presumably, neighbor as well.

For love of neighbor to truly flourish, we must have some underlying awareness of the ontological and metaphysical connection we have with one another. It cannot flourish based on a moral psychology of possessive individualism and selfishness, however much one might argue that "to appear loving" is of reputational value. Unfortunately, when it becomes disadvantageous for the disconnected individual to "love thy neighbor," then he or she will do whatever is selfishly useful as long as it can go undetected. "Contractarian" self-interestedness can construct certain minimalistic prohibitions against such things as wanton violence and lying simply because we need these to live at all successfully together. But once those prohibitions are no longer advantageous to the greedy individual, and the power of the Leviathan (i.e., the police state) weakens, a state of nature breaks out that the philosopher Thomas Hobbes captured with the phrase, "a war of all against all."

So if we cannot provide a foundation for a universal ethics based on contractarian psychology, or reason alone, or compassion, then what? In the final analysis, we absolutely need the underlying connective tissue of the One Mind in us all. As Henri Bergson wrote in 1932 in *The Two Sources of Morality and Religion* (English ed., 1935), nothing can compare in moral power "with the feeling which certain souls have that they are the instruments of God who loves all men with an equal love, and who bids them to love each other" (1932/1935, 311). He wrote of something that Sir John would easily have agreed with based on the concept that all our small minds are part of a Divine Mind—"the mutual overlapping of psychic states" (1932/1935, 313). Bergson, in a statement on joy that Sir John would also have embraced, wrote, "Joy indeed would be that simplicity of life diffused throughout the world by an ever-spreading mystic intuition; joy, too, that which would automatically follow a vision of the life beyond attained through the furtherance of scientific experiment." (1932/1935, 317)

Sir John captured one important dimension of nonlocal mind by emphasizing that thoughts have a certain energetic reality about them that we do not yet understand scientifically, but the thoughts of an individual can travel through the larger Infinite Mind and impact others. This topic gets very metaphysical, of course, but it is a core element in the Unity school of Christianity and in New Thought generally:

> Our thoughts are, most assuredly, things. They are conceived
> in the mind and travel through time and space like ripples in
> a pond, affecting all they touch. (2012, 4-5)

Sir John cited a phrase from Charles Fillmore, "Thoughts are things" (2012, 20) with frequency. In other words, while our thoughts create the blueprints of our actions and are in the obvious sense our building blocks, for everything ever created by any human being begins as a thought, Sir John is going a little further than this and giving thoughts a deeper and even more significant dimensionality.

If God or Ultimate Reality is Infinite Intelligence or Divine Thought, then maybe the entire universe emanates from thought. Sir John would now and again quote Ernest Holmes (1887–1960), founder of the worldwide Religious Science movement that is one of the three pillars of New Thought. In his classic work *The Science of Mind*, Holmes wrote a section titled "Thoughts Are Things," taking this from Fillmore, as did Sir John. Holmes stated, "Thoughts are things, they have the power to objectify themselves; thought lays hold of Causation and forms real Substance." (1938/1998, 144)

Sir John wondered if Thought in the form of Divine Mind is the source of all that is and the sustaining attentive presence that holds everything in existence continually. He wrote this very powerful statement, thoroughly capturing the content in the previous paragraph:

Thoughts are things. Thoughts create things. Thoughts shape things. Thoughts are real. The invisible processes going on inside our heads that we call thinking produces objects as real as the ground we walk on or the food we eat. Consider for a moment your thoughts as a flowing mountain stream. The life-giving stream begins high in the mountain, flows down to the valley below, and then empties into the fields and orchards of your life. (2012, 20)

Thoughts are things. They become things. And Sir John thought we potentially would understand how the energy of Divine Idea can transpose from the key of the invisible spiritual Godhead into the energy and material of the universe.

REACHING HARMONIC RESONANCE WITH UNLIMITED LOVE

The disciplined human mind as Sir John understood it can focus and concentrate on the idea that all people are equally in God's image and equally worthy of love, and thereby extend love more and more widely to the point of equal regard. Moreover, in slowly approaching this inclusivity, Sir John felt that we might come into a kind of harmonic resonance with the energy of God's Unlimited Love and experience a mysterious uplifting spiritual mind that moves us well beyond whatever limited love the human substrate is capable of on its own.

I am reminded of my strenuous efforts as a surfer at Gilgo Beach trying to catch a wave and then being swept forward by its astonishing power and speed. Our most concentrated, 100 percent effort to love everyone is actually only very tiny (symbolically, just a percent or two) compared to the massive energy of the love wave that we might catch and ride. Another useful analogy is an airplane that must achieve some minimum speed on the ground under its own power before the air pressure becomes greater under the wing than over it (Bernoulli's effect), and the plane lifts suddenly off into the sky at rapid speed. Sir John wrote,

As we learn to love more, do we engage a higher frequency range of our hearts' qualities and quicken our intuitive connection with Spirit? Could unlimited love be described, in one aspect, as a cosmic energy in which human participation is possible? Does such love produce health and peace in our world and aid and participate in humanity's purpose? (2003, 94)

In surfing or flying, of course, the efforts exerted by the human arm or the engine are proportionately minimal in comparison to the power and energy of a wave or of air pressure. Sir John is inviting us to consider the possibility of a much more intense love energy—something millions and even billions of times greater than anything humans are capable of on their own—that may find us and surprise us out of nowhere when we make our best effort to love people without exception, even when they are otherwise unlovable. Actually, Sir John felt that if we look hard enough, we can find some precious aspect of God's image in every human being without exception. In other words, he was interested in how those who put their fullest effort into a virtuous loving life may, on the basis of this condition, find a window into the Divine that is equally a window into inner peace.

Sir John wrote, "Often in our ignorance and nonthinking, we hinder God and stop the current of divine messages. But when our lives are permeated with a lively faith and a sincere desire to learn, messages of love and guidance flow to us like a beautiful river that has found smooth passage through our life stream" (2012, 78). In other words, at the metaphysical ("What is?") level, Sir John understood at least some aspect of our minds to be eternal and part of some form of collective unconscious Mind that includes God. At the epistemological ("How do we know?") level, he therefore affirmed that our minds are conduits for divine inspiration and inspired insights that come from God. In a very interesting passage he continued this theme:

It isn't by trying to squeeze the required thought out of our brain cells that we get the knowledge we need. Rather, it is by opening up our minds, freeing, letting go, expanding, and broadening our thoughts with the implicit understanding that we may become more open to the infinite mind of God in universal form. Who knows what realms of knowledge and learning abide therein, waiting for us to be receptive. (2012, 78–79, italics added)

This quote is fascinating because it vividly indicates Sir John's opinion that our minds and our best thoughts may be gifts of knowledge and creative insight from "the infinite mind of God." His comments on listening as "the way we open ourselves to another's point of view" pertain not just to human interactions but to the encounter with God. (2012, 77)

His interest in creativity was shaped by this view of inspired knowledge, consistent with theologian Karl Barth's writing on Mozart or Michael Faraday's self-reported "seeing" of electromagnetic fields. Of course, in modernity, the idea that we receive knowledge from a higher intelligence in which we can participate if we are open to it is almost antithetical to empiricism, but for Sir John what we receive can be examined and reexamined by the scientific method in the form of questions, hypotheses, and predictors.

Against the above metaphysic of mind and theory of epistemology, would anyone be surprised to find that Sir John believed that if we live in generosity and virtue, we can actually get to a point where we might more easily receive and absorb the light of God's Unlimited Love? Growth in Unlimited Love is to move toward God: "*As we learn to love more, do we engage a higher frequency range of our hearts' qualities and quicken our intuitive connection with Spirit? Could unlimited love be described, in one aspect, as a cosmic energy in which human participation is possible? Does such love produce health and peace in our world and aid and participate in humanity's purpose?*" (2003, 94). Here we are reminded of the ancient idea of a "ladder of love," in which by using all of our abilities to love others, we reach a level of attunement to love that allows us to at last connect with the cosmic energy of Unlimited Love that many people otherwise call "God." Love may even be an energy that holds

the whole world together like a cosmic glue: "Those who are philosophically inclined may find it helpful to understand God's unlimited love as
the original and ongoing basic creative force of the universe. This love
was present before the beginning, and it continues to hold all things
together. Our fleeting human emotions and perceptions are in fact mere
glimpses of God's perfect love." (2000b, 19)

Consider the sober poet W. H. Auden, my favorite, who described his
quiet experience of agape love as follows:

> One fine summer night in June 1933 I was sitting on a lawn
> after dinner with three colleagues, two women and one man.
> We liked each other well enough. But we were certainly not
> intimate friends, nor had anyone of us a sexual interest in
> another. Incidentally, we had not drunk any alcohol. We were
> talking casually about everyday matters when, quite suddenly
> and unexpectedly, something happened. I felt myself invaded
> by a power which, though I consented to it, was irresistible
> and certainly not mine. For the first time in my life I knew
> exactly—because, thanks to the power, I was doing it—what
> it means to love one's neighbor as oneself. I was also certain,
> though the conversation continued to be perfectly ordinary,
> that my three colleagues were having the same experience.
> (In the case of one of them, I was able later to confirm this.)
> My personal feelings towards them were unchanged—they
> were still colleagues, not intimate friends—but I felt their exis-
> tence as themselves to be of infinite value and rejoiced in it.
> (1965, 30–31)

Auden captures an experience that is subtle and deeply emotional, an
Unlimited Love animating and enlivening his sense of awe for the gift
of others. Was this experience unconditional? It appears to be so, as a
deep sense of the overwhelming presence of Unlimited Love simply
comes out of nowhere, as it often can. So yes, God's love is ultimately
unconditional, but we still should be doing everything we can as active

agents to abide in it as a law of life, either before or after such a dramatic experience. Auden's experience is dramatic, however, in a rather quiet way. He does not fall down on the ground in a physically violent, seizurelike event. Some people do so, and that is fine, but in Auden's case we have someone whose experience is more of a surprising deepened awareness, one consistent with the tradition of the Anglican contemplative mystics, such as Auden's contemporary Evelyn Underhill. Sir John, who greatly appreciated the Anglican branch of Christianity and contributed significantly to the reconstruction of a major room in London's Westminster Cathedral, would likely resonate with Auden's experience of Unlimited Love quite closely.

Let's return to this passage from Sir John: "As we learn to love more, do we engage a higher frequency range of our hearts' qualities and quicken our intuitive connection with Spirit?" (2003, 94). My guess is that Auden, a devoted Anglican whose poetry was clearly flowing from the hand of someone who took God's love with seriousness, was at least not likely to abuse or trivialize his reported experience of Unlimited Love.

Many people assume an ideal quality of love that cannot be known apart from God's presence. It is perfection in love that can only be born from above, from the Spirit. The quality of love born from the human heart alone is a magnificent hint at God's love, but still it is changing and unstable, often exclusionary, lacking in wisdom, and rather routinely overwhelmed by such things as fear, envy, greed, bitterness, and revenge. How could anyone doubt the limitations of human love? So many of us live in the pursuit of power, fame, or possession; our love is thus more manipulative than pure. Can even the most optimistic observer of human nature doubt this? Of course, human love at its best in a particular time frame can look magnificent, but it will eventually show its flaws. If it were not for the frequently reported experiences of a God of perfect love whose creative presence shapes and enhances our human substrate, the world would be dark indeed. God's love is unchanging, available to all, perfectly wise (and therefore we need to be patient with it), and never overwhelmed.

This spiritual love comes from the Divine Mind that touches our spiritual mind within. We do not get to Unlimited Love by cold logic. Sir John recognized the power of the mind in extending love unconditionally as a matter of focus and principle, as well as the importance of good role models. Thus he greatly adored the rational mind. But he also took seriously the idea of Unlimited Love breaking into consciousness as an energy that might take us by surprise. So he wrote, "*The divine spirit may move in your life and make it over from within so that you see things in a new light, and love may become the spontaneous expression of a Spirit-filled soul.*" (2012, 76)

Referring to Rufus Jones, the famous Quaker mystic of the mid-twentieth century, Sir John writes,

> Rufus Matthew Jones taught that those who worship God are empowered by the Spirit and that religion is not a burden but rather a means of being lifted up to new heights of joy and philanthropic achievement. The divine spirit moves into your life and makes it over from within so that all things are seen in a new light, and love for all becomes the spontaneous expression of a spirit-filled soul. (2000a, 127)

Sir John was a Tennessee mystic:

> God is said to be the source of love. Love cannot flow in unless is also flows out. Does the spirit of god resemble a stream of water and are god's followers like many beautiful fountains fed by this river of water? (2000a, 127)

Here is the image of the perennial philosophy in which the Godhead overflows with love and in which we can participate. But Sir John would view this as conditional on our doing everything we can with the power of our God-given minds to extend love to all people in an equal-regarding manner. Sir John clearly emphasized a strenuous and disciplined self-control of one's mind as a way to "crowd out" bitter, hostile, and unkind thoughts by cultivating loving ones. The mind has to be filled with some kind of thought, and if we exert control, we can grow in love through the power of mental focus and concentration on loving ideas. Then we can be surprised by the direct experience of Unlimited Love, or be awakened to the already existing unity between our individual minds and the Divine Mind.

THE LIMITS OF ALL MERELY HUMAN LOVE REALISTICALLY CONSIDERED

For all its strength, human love is astonishingly weak. Going through dim periods with hardly a bit of energy, human love is easily overwhelmed by adverse social norms, group pressures, personal disappointment and bitterness, envy, hatred, anger, vengefulness, greed, and the will-to-power. For health-care staff, the stresses of a long morning commute on the expressway, the anxieties of budget cuts and economic uncertainty, and being treated callously by doctors or colleagues can all contribute to an uncaring day, especially as the end of the shift rolls around.

For all its wise aspirations to do good, human love is astonishingly unwise. Some say it is better to teach people to fish than to give them one, and yet nothing is better than without hesitation giving a fish to a hungry human being. Parents love their children, yet parents can overindulge children so much that their lives are completely ruined for lack of self-control, purpose, character, and responsibility. We seek to help a friend stay true to his core values and we end up being an obstacle to his spiritual growth. We think we know objectively what will bring someone happiness but it turns out to cause misery. Love requires wisdom and a respect for the freedom and autonomy of others in their decisions, though without abandoning them or failing to offer recommendations.

For all its purity, human love is astonishingly impure. The cynics are wrong in asserting that every look of kindness is intended to manipulate, but the skeptics are right in thinking that sometimes this statement can clearly be true. Even parental love at its sacrificial best can be a possessive effort to overcome one's own disappointments through the success of the child. All human love has benefits for those who bestow it, at least in the form of meaning and gratification, and often in the form of deepened relationships and joy. But it can and does rise above reciprocal calculations and reputational interests. Still, we often see the law of reciprocity limit human love. "I don't do nothing for nothing," or so they say. Human love can be amazingly pure. We know people who selflessly model generosity and kindness with no greater hope than that others might be inspired to go and do likewise. But let us acknowledge that sometimes purported love is just a thin veneer over self-interest.

For all its inclusiveness, human love is astonishingly exclusive. We claim a noble love of all humanity without exception, and yet many of us love only those who are most like us. Perhaps we believe in political diversity, but when people disagree we become uncivil and even demonize them with hateful words. Of course, we tend not to demonstrate much self-awareness when it comes to our biases, and we frequently deny that they exist when others clearly see them. We see those who profess universal love and respect caught up in class snobbishness. Philanthropy at its best rings true with a passion for a shared humanity, and at its worst is a dry action of condescending tolerance. Ardent secularists can claim to love humanity but dismiss the thoughtfully spiritual, while religious fanatics whose traditions teach of God's unlimited love can arrogantly attempt to annihilate the unbelievers. When leaders like Gandhi, Rabin, Sadat, King, and Lincoln assert the equal value of all regardless of religion or color, they are assassinated. Many good neighbors in Germany, often churchgoing people, were quickly able to tolerate the Nazi eradication of the Jews.

Our human nature is capable of love, but when we have an honest moment of sincere self-assessment, we know that we are merely human. Any honest moral inventory establishes without exception that each of us falls well short of the glory of Perfect Love. In general, the perfect is the enemy of the good, and we should not diminish the significance of the love we have for the Perfect Love that we do not have and never will. We oscillate between love and hate, with a large space for wanton indifference between the two. Sometimes we treat the very loving as saints, but often not until we have killed them for shedding light on the darkness within us.

While any human love may have its ultimate ground in God's love, human love is fickle and unstable. It can be stabilized and extended by the power of the mind and right thinking. But more is needed. Human love must be completed or perfected or made whole by God's love acting upon the human substrate, through the spiritual mind.

Chapter Six

Why Are We Created?

LIFE AS AN OPPORTUNITY TO OVERCOME HARDSHIPS, TO WIN
WITH LOVE, AND TO ACHIEVE JOY IN GOD FOR ETERNITY

SIR JOHN ASKS, "Does a divine fountainhead of love exist in the universe in which degrees of human participation are possible?" (2003, 87). He believed so, and herein was purpose. Such participation is the ultimate purpose of every human life. All else is secondary and instrumental.

In Sir John's theophilosophy of purpose, God wants each of us to participate as co-creators in God's creative loving purposes, to love God and to love neighbor as self, and from this we can experience true joy. Because each of us has a tiny part of the eternal Divine Mind within us, we are creatures through whom God realizes God's own creative and love potential. Our spiritual mind is identical with the source from which it has been given, the Divine Mind. We are each one with God, as all the mystics who have looked within—whether Christian or Jew, Hindu or Muslim—have attested. They all speak of the ineffable experience of the soul being uplifted into the Infinite. Our purpose is to serve as the manifestations of God's Mind in loving, co-creative purposes.

Sir John understood that to be joyful in life, we need to know that there is a reason for us to be alive, that our lives matter, and that we have a purpose. The human need for purpose is obvious, as no one flourishes without it. We rather flounder, somewhat like the Greeks' gods and goddesses who seem to have nothing much to do other than bicker with one another. The major spiritual problem people grapple with is the sense that life is without purpose.

The pursuit of a noble purpose allows us to grow, and life on earth is about growth. Allow me to be a bit pastoral, because Sir John was that way. Our lives always involve deaths and resurrections, large and small. Sir John experienced those pains personally with the loss of his first and then his second wife; of his only daughter, Anne Templeton Zimmerman; and of all the dear friends who pass away before the eyes of anyone who lives to be ninety-six years of age. Life as we have known it in a certain way always ends. There might be a town we loved, a job that meant everything, a relationship that made getting up in the morning a joy—everything at some point changes. In these changes we grow spiritually as God takes us to a new life. God never takes us backward but only forward in the newness of life. The old days were probably not that perfect anyway. For Sir John, life was all about growth, becoming, process, change, resurrections, and transformations. He did not wish to hear any negative messages from the pulpit—or from anywhere. Life is lived "implaced," as the philosophers write, but it is also always a journey to newness in which we had better not become overly attached to the way things used to be. God loves us no matter what and calls us to the newness of life and growth through all the deaths and resurrections of life. The peak ahead is higher than the valley of the moment. But to flourish in such a journey we all need a purpose and a desire to grow. Purpose is the key human spiritual problem; Unlimited Love is the answer.

The first Templeton Prize was bestowed upon Mother Teresa. Sir John admired her greatly and noted that she was concerned about the absence of purpose in the lives of many people in more affluent nations:

> Mother Teresa reported in some of her television interviews that she finds the greatest poverty and desolation among the wealthy of the world today. She noted that there is a stark need for mission to offer love and nurturing to the barren of heart. (2012, 174)

Young people in industrialized nations struggle with anxiety disorders, depression, high suicide rates, alcohol and substance abuse, and a general existential malaise. As Christian Smith has shown along with others, as of 2010 American youth are on the whole more narcissistic than they were in earlier decades. How can youth find a nobility of purpose? How can spiritual growth be awakened?

LIFE ON EARTH AS PREPARATION FOR ETERNITY

Sir John did not shy away from finding meaning and purpose in our travails and sufferings in this world:

> If earth is a school for infinite purpose, wouldn't a likely question be: Who are the teachers? Life is often filled with joy and sadness, challenges and overcoming, successes and difficulties—all of which offer opportunities for learning and progress. Can we perceive how these ranges of thoughts and feelings, situations and experiences play a constructive teaching role in our lives? From a divine perspective, would our soul's education be incomplete without these experiences? (2003, xv)

Here we see no anger at God for life's hardships, any need to assert that life is intended to be a smooth ride, or that something is terribly wrong with suffering. No, Sir John simply accepted the peaks and valleys as part of the territory of life; he would agree with the simple statement, "Hard lessons are learned hard," and life is all about learning those lessons and about deep growth. Trials and crises are sometimes desperately needed if a person is going to forget about one's own routinized customary self and become new, something different.

Sir John considered life here on earth as preparation for eternal life and an opportunity for growth:

> The Christian religion often speaks about the way in which the human experience on earth is to function as preparation. To progress spirituality can be to increase our love of God, our understanding of God, and love for his children. Although our

body has a physical reality, can it be only a temporary dwelling place? Our physical body will someday die. But maybe death destroys only that fit for destruction? (2000a, 132)

In this line of thought he again is consistent with the perennial philosophy.

In our journey here in the physical world, adversity may well be our most effective teacher:

> Therefore, one teacher may be *adversity*. Could it be that adversity can play a constructive role in life? Could it be that without these experiences our education would be incomplete? Maybe from the divine perspective, can the sorrows and tribulations of humans help to educate us for eternity? (2000a, 133)

Because trial and self-discipline are so important to growth, Sir John titled the relevant chapter in *Possibilities*, "Can Earth Serve as a School?" (2000, chap. 10). He did not think that life is easy, devoid of major obstacles or failures. The key to growth is persevering and staying on course despite opposition and disappointment. A chapter of *The Humble Approach*, written years before *Possibilities*, is titled "Earth as a School" (1981/1995, 90), which indicates that this was a core idea for Sir John over many years. The purpose of life on earth is to grow in spiritual depth. We are here to grow in love, patience, wisdom, and understanding. Sir John asks elsewhere, "Is our material world a possible incubator, provided by the Creator, in which human spirits can develop and seek ultimate expression in a realm invisible to us?" (2003, 23).

And what specifically defines growth? What outcome is ultimately a measure of growth? Sir John wrote,

> Why should we waste even one day? At the end of each day, can we say we have learned to radiate to all pure unlimited love, and helped our neighbor to learn the joy of giving love? (2000a, 136)

With regard to eternal life, he questions boldly,

> "What grand opportunities abound for unlimited love and humility to open up the doors to exciting research into the science of the soul? Who knows, the spirit of unlimited love and humility, which we build during our lifetimes, may be the part of the human person that is immortal! Radiant, unlimited love, universal, eternal!" (2003, 92)

Obviously Sir John was not a man to shy away from bold thinking consistent with the mystical and perennial tradition. He was never willing to accept the standard assumptions of materialism as the last word when it comes to human nature. He held out for the eternal soul and its destiny in the invisible dimensionality of being.

Sir John stayed true to this idea of spiritual growth, which he had earlier defined as "overcoming our ignorance and self-centeredness until we are in tune with the divine" (1981/1995, 95). Furthermore, he wrote,

> Of course all of us should work for self-improvement by prayer, worship, study, and meditation. But one of the laws of the spirit seems to be that self-improvement comes mainly from trying to help others—especially from trying to help others to enjoy spiritual growth. Growth comes by humbly seeking to be a more useful tool in God's hands. Giving to others material things helps the growth of the giver, but often injures the receiver. It is better to help the receiver to find ways to grow spiritually himself. (1981/1995, 95)

This growth seems inevitable, for "once a person is captivated by *agape*, there is no turning back" (2000b, 41–42).

The closer we get to Unlimited Love, a love for all persons equally, the closer we are to participating in God: "Embracing *agape* does not make us God, but the Bible suggests that, unlike other forms of love, it allow us in some way to participate in divinity" (2000b, 19). "If we release egotistical self-will," he writes, "and invite the richness of divine love into our lives, do we become more effective channels for the Creator's love and wisdom?" (2003, 91). Sir John writes, "Scripture tells us, 'It is better to give than to receive' (Acts 20:35). How can the talents, abilities, intelligence, and success with which we are blessed be returned to the world in some form that will benefit humanity? Unselfish giving can be a sign of personal and spiritual maturity." (2003, 91)

Who are the teachers? "One teacher is adversity," Sir John explains (1981/1995, 93). "Growth can come through trial and self-discipline. There is a wealth of evidence indicating that too much prosperity without work weakens character and causes us to become self-centered rather than God-centered. Spiritual growth and happiness do not come from getting but from learning to give," writes Sir John (1981/1995, 93). He continues,

> How could a soul understand divine joy or be thankful for heaven if it had not previously experienced earth? How could a soul comprehend the joy of surrender to God's will, if it had never witnessed the hell made on earth by trying to rely on self-will or to rely on another frail human or on a soulless man-made government? (1981/1995, 93)

Moreover, "It is apparent that sometimes a great soul does not develop until that person has gone through some great tragedy. Let us humbly admit that God knows best how to build a soul" (1981/1995,). Finally, "As a furnace purifies gold, so may life purify souls. When a man is born into the world, he is like a piece of charcoal" (1981/1995, 94). One of the things I liked most about Sir John was his deep optimism about human spiritual potential as God-infused, and his equally deep pessimism about human nature on its own divorced from a relationship with God.

Sir John's view of earthly life as an opportunity for spiritual growth in love is clearly grounded in a view of the nature of human consciousness. If we are merely material beings who are entirely extinguished upon death, then this idea of spiritual growth as the purpose of life makes no sense. Sir John always saw this life as preparation for something beyond time and space, and as a fleeting opportunity to develop toward Unlimited Love. In the perennial philosophy tradition, he could thus envision life in this visible world as a school for the growth and development of our spiritual and eternal natures. The more wisdom, love, forgiveness, and gratitude we develop in this visible world, the better off we will be in the invisible world.

Our existence in the material world is a transitory state. For Sir John, mind and consciousness have a primacy that a materialist metaphysic cannot contain. We are all connected with Divine Mind, and something about each of us is unlimited, some aspect of ourselves that transcends space, time, and the material world. For Sir John, the central dogma that the mind cannot exist independently of the brain was open to question, as is the dogma that all the "just right" physical laws and fine-tuned constants of the universe that give rise to life are not reflections of an Infinite Mind with purpose behind it all.

Yet in a larger sense, all human purpose for Sir John exists under a canopy of a loving God who is a conscious intelligence behind the universe from which the ordering principles of the universe spring, the foundation of the universe itself. He wrote often of the universe as "fine-tuned" to give rise to an evolutionary process. He loved the visionary physicists, to whom we turn in part 2.

JOY AS THE BY-PRODUCT OF THE LOVE OF GOD AND OF NEIGHBOR AS SELF, DESPITE THE INEVITABLE HARDSHIPS AND FRAILTIES OF OUR FLEETING LIVES ON EARTH

Sir John wrote, "The ability to choose to move to a higher spiritual level of consciousness and to a happier, more fulfilling way of life lies entirely within each individual. A Universal intelligence flows through us." (2003, xix)

There is no separating joy from the unitive love-knowledge of God and from contributing to the lives of others. Joy has two sources in his thinking: the love of God and the love of neighbor as self.

Life is never too easy, but it helps to be joyful. Yet people so often seek joy and never find it, because they are seeking it directly or superficially. Sir John felt that the best way to find joy is to allow it to emerge as a by-product of contributing in concern, care, and love to the lives of others, and of loving and being loved by God. As a Tennessee-born Presbyterian, he affirmed that joy flows from sincere love for God and neighbor. Joy is not the primary motivation of sincere love, but the odds are high that it finds us even if we are not seeking it. Sir John wrote,

> Who are the happiest people you have ever met? Let us write down the names of ten persons who continually bubble over with happiness, and we will probably find that most are men and women who radiate love for everyone. They are happy deep inside themselves because they are growing spiritually and fulfilling God's laws. Jesus said, "Thou shalt love the Lord thy God with all thy heart, and with all thy soul and with all thy mind. This is the first great commandment. And the second is like unto it, Thou shalt love thy neighbor as thyself." (1981/1995, 98–99)

Joy flows not from self-love alone, but rather from the love of self that is interwoven with sincere love for God and neighbor—the three conjoined in a tapestry. Within so many spiritual systems, deep joy is generated by an awareness of God's surprising and overwhelming affirmation of our very being and that of our neighbors. Sir John had nothing at all against the love of self, but he endorsed only the right and most enduring self-love that comes when we discover the joy of abiding in the double love commandment (Matthew 22:37–40) that he often cited, as in the above passage. Sir John took good care of his body and mind, which he considered to be God's special gift and responsibility. He lived to be a lucid ninety-six-year-old in large part because the love of God and neighbor gave him a reason to be a good steward of his health based on meaning and purpose. He liked to smile and made a point of it. He devoted himself diligently even in his last years to developing a vision about how to help all humanity through supporting innovative spiritual progress that in the long run would do the most good.

Now, why does Infinite Mind share its essence with us? Why does God breathe a part of this essence into us, making each of us more than a material being comprising biochemical reactions? So that we might be co-creators with these minds, and so that we might know joy in loving God and being loved by God. Our purpose is to love God and to practice love, "and to find something to appreciate in every person we meet" (2012, 177). Giving to others in helping actions, creativity, and philanthropy "has to do with our health, happiness, and overall well-being. Does your being not sing when you give from the love within your mind and heart?" (2012, 181). As Sir John often wrote, "Happiness is a by-product." It is not found by seeking it, but it is "a by-product of trying to help others" (2012, 255). After all, each and every one of us is equally loved and equally participant in God.

Maturity, states Sir John, is to "feel a surge of joy over the good fortune of others" (1987, 147). This joy is deeper than mere happiness, which is contingent on external circumstances. Sir John has in mind something that endures under any conditions. But in an even deeper sense, joy is a fruit of being open to "seeing nature and reality as that which God speaks through" (Templeton and Herrmann 1989/1998, 38). Here he quotes the seventeenth-century British metaphysical poet Thomas Traherne:

> Your enjoyment of the world is never right till every morning you awake in Heaven; see yourself in your Father's palace; and look upon the skies, the earth, and the air as celestial Joys; having such a reverend esteem of all, as if you were among the Angels. . . . You never enjoy the world aright . . . till your spirit filleth the whole world, and the stars are your jewels; till you are as familiar with the ways of God in all Ages as with your walk and table: till you are intimately acquainted with that shady nothing out of which the world was made; till you love men so as to desire their happiness, with a thirst equal to

the zeal of your own; till you delight in God for being good
at all: you never enjoy the world.

Here one sees that joy is an exuberance born of a relationship with God
and with the universe as God's gift to us all.

Sir John found joy in God's gift of nature and the universe, and in
God's imminent and totally loving presence in our minds and at every
level of the created order. He was a palpably joyful man, and joy was
one emotion that he wrote of extensively. Joy is the state of being we
arrive at when we love God and love neighbor as self with sincerity, and
when we celebrate the marvelous creation. He also understood joy in
the Pauline sense as a "fruit of the spirit." The poet and Nobel Prize
winner Czaslaw Milosz captures this idea as follows:

In advanced age, my health worsening, I woke up in the mid-
dle of the night, and experienced a feeling of happiness so
intense and perfect that in all my life I had only felt its pre-
monition. And there was no reason for it. (2003, 70)

Joy as Sir John understood it is a gift from a loving God, and joy really
is God's hope for our lives as we grow in love and metaphysical insights
into the reality of God as the Ultimate Reality sustaining all that is.

An astonishing joyful picture of Sir John shows him with his arms
extended out while standing in the surf on a bright sunny day in Nassau, with a smile a mile wide. He celebrated God's love and being made
in God's image in a joy so lively that it jumps right off the photo and
into one's soul. He was never loud, but in a deeper sense he was always
shouting "Hallelujah" and praising the wonderful universe and our
minds within. This was his spirituality. He heard God's music at every
level of the creation, and in every detail. He celebrated the universe
from the blue ocean waves to the skies above his modest beachfront
home in Nassau.

Sir John considered it impossible to feel dull and unhappy when
reaching out to do good to others. Those who lose the joy of living are

the ones who cannot think beyond themselves. The toxic thinking that colors life in shadows gets totally out of control when we become pre-occupied with the self and its resentments. Our emotions turn sour. The most helpful thing is to cast off negative thoughts and related emotions by actively and personally reaching out and helping people in ways that they desire and appreciate. Positive emotions of joy, gratitude, and hope will generally follow right along unless the recipient is hostile to what we are doing, in which case it is time to move on. Our hearts become more radiant when we help those who want our help.

When we give to others without thinking of self, we set free in simple practice a dynamic of the human personality that is the single most essential element in a joyful life. Sustainable joy is primarily the result of contributing to the lives of others. Of course, money can make us happier when it makes life easier—for example, it buys us health care, a peaceful neighborhood, a decent place to live, healthy diet, good schools, and a stable job—or when we are able to give it away to help others. Decent circumstances do matter when it comes to a satisfying life. But wealth does not ultimately determine whether we are miserable or joyful. The "hedonic treadmill" of comparison with the wealth of others is a term that psychologists Philip Brickman and Donald Campbell coined in 1978. They found that financial windfalls are no ticket to heaven, and sometimes they create misery.

The most obvious path to joy lies in consistent small actions that brighten up those around us; as a by-product, we shift our own emotional state in the direction of joy. In some deeply contented moments we feel uplifted because the life of someone else comes to mean as much or even more to us than our own. The tension between self and other evaporates, and we awaken into a new world where the distinction between my good and another's fades. Joy just arises out of this love, like it does for a child absorbed in play. We never find joys when we settle for selfishness. In this sense, it really is better to give than to receive, or at least just as good.

Love does not allow us to hold on to our miseries. Love is like the sun, and we are like sunflowers. The more we are turned toward love, the more we live abundantly. A passage from Paul means much: "God loves a cheerful giver" (2 Corinthians 9:7). This love is the glue that binds us together with others in the communities that make life joyful.

A joyful life revolves around at least one immaterial good: love. We cannot grasp love like a coin, but this warmth and concern for another are more real and meaningful than anything we can possess. Here is an exercise: Close your eyes and intensely imagine giving a generous smile to the person in your life you love most; then open your eyes and you will feel your heart strangely warmed. Deeper happiness, which is an enduring and unselfish joy, does not lie in worldly power and fame, although a good life will hopefully be recognized and celebrated as such; nor does it lie in extra riches that are divorced from creative giving. We all have real needs for tangible basics, and having the basics relieves stress. Insecurity about food, shelter, and health care are obviously stressful and unhealthy. But ample research shows that deep joy does not come from that fancier car or even a second home. These external successes aren't all bad, but they are fleeting victories and at best have surprisingly limited returns. We may fill up the closet or the house with more and more, only to discover that our lives feel emptier and emptier. The best way to enjoy life is to loosen our grasp and be freer to love.

No one who is feeling anger or fear, for example, can simultaneously feel joyful. But when we love others, we move into a state of internal harmony that is also a state of deep joy. Washington Irving understood this inner joy when he wrote, "Love is never lost. If not reciprocated, it will flow back and soften and purify the heart." Simply to abide in a spirit of love is to abide in a joyful warmth and harmony with oneself. Love is its own emotional and spiritual reward that no one can take away—even in the absence of reciprocation. Of course, love often does beget love, just as hate usually begets hate. We often find joy in the deeper relationships and connections that love brings into our social lives. But we cannot count on reciprocity and should not depend on it or even seek it. We have only to love people and hope that they will be inspired to "go and do likewise."

Simplicity

There is wisdom in the old Shaker hymn, "'Tis a gift to be simple, 'tis a gift to be free." In a culture that always has us living like kids in a candy shop, we need self-control and a simplicity because otherwise we are in for every kind of disordered love of countless little things that will never make us truly joyful. We need temperance and moderation, a freedom from comparison with what others possess, and a willingness to concentrate on relationships rather than the accumulation of material things. The Roman philosopher Marcus Aurelius wrote, "Remember that very little is needed to make a happy life."

Simplicity allows us to find delight in the beauty of life that is available to us all—the waves of the ocean along the shore, a cool breeze on a summer afternoon, the warm glow of an outdoor fire, a quiet moment of companionship. The simple gifts of life are infinitely wondrous if only we can appreciate them in their glory and be grateful. The red and gold of autumn leaves, cloud and sun, ice and snow, shells and streams, trees in spring, fireside conversation, meals eaten together, games in the open air, the satisfaction of a lute suite well played in the evening, the simple dedication to learning and truth—all things plain yet elegant. Simplicity allows us to celebrate nature, to be attuned to the generative planet earth on which we live. In simplicity is the wonder of a child, gladness about the newness of life each day.

The love Sir John wrote of is unlimited, unconditional, active, and joyful: "Unconditional love. Unlimited love. Active love. Joyful love. The option to grow in *agape* is open to everyone on earth. It is an invitation to true happiness for you and others. May it become our aspiration, our expression of God's love, radiating through us" (1999, 101). One must note, however, that this joy is not a primary or even a direct motivation. It is secondary, or even a mere unintended by-product of love, an aftereffect, something unsought. But it comes with the territory of love. Sir John noted, "As a friend of mine is fond of saying, '*Love given is love received*,' and happiness is a by-product of that kind of love. This law of life, when put to use through loving expressions and prayer, can guide us in fulfilling our every aspiration, as well as enriching the lives of those around us." (1997, 242)

Perseverance

While Sir John acknowledged the lives in which people are called to love under difficult circumstances that may lead to challenges of many kinds, he saw that, more generally, abundance and flourishing follow from the life of love when we stick with it: "Although for some, a commitment to agape has entailed martyrdom, to focus on that is to miss the over-whelmingly positive meanings of the word. Agape is active love, love that reaches out to others. It is joyful love, offered not out of obligation, but in a spirit of compassion and hope" (1999, 100). People of great love do not always have an easy time of it, and they may be met with rejection. However, "In the consciousness of love, one is able to ignore adversity, insults, loss, and injustice" (2000b, 47). "Remember an important truth: whatever the need or circumstance, love can find a way to adjust, heal, or resolve any problem or situation." (2000b, 56)

Sir John saw the suffering that human beings inflict on one another each day. The betrayal, rejection, and contempt that are characteristic of human beings never surprised him, and he never thought that these elements could somehow be removed from the human condition. He experienced considerable suffering in his life, including the death of both his first and second wives, and his beloved only daughter. He understood that no matter how fortunate we may be in life, we are all going to pass on, and most likely death will come with decline, disease, and possible pain. Perhaps worse, every single person we love will suffer the same fate with time.

There is plenty of frank acknowledgment of suffering in Sir John's writing. He understood well the Christian theme that doing the right thing often causes some degree of misunderstanding and produces some suffering. Other people expect goodness to be rewarded with trophies in one form or another. They sometimes lose faith and turn bitter when they encounter rejection and pain. But visionaries grow in faith during the desolate times. Other people perceive all pain as an evil waste. Sir John clearly saw hardship as an avenue for growth.

There is no greater visionary than the prophet who is responsible for the central chapters of Isaiah. This ancient seer saw that suffering in the cause of God could lead to healing and liberation. More than any other among the great prophetic spokesmen, this Isaiah sees further into the redemptive nature of God and accepts the suffering of God's true servants. The more godlike a believer becomes, the more likely some persecution will come from those of little or no faith. Isaiah includes some wonderfully poetic passages filled with hope and joy for the new age that God will one day bring on earth, a beautiful time when injustices and hatred, violence and war, will be no more:

> "The wolf shall dwell with the lamb and the leopard shall lie down with the kid. The calf and the lion shall play together and a little child shall lead them. They shall not hurt or destroy in all my holy mountain; For the earth shall be full of the knowledge of God as the waters cover the beds of the sea." (Isaiah 11:6–9)

How do we ever get to that marvelous future? The central chapters of Isaiah see it happening through a unique servant of God. This servant shall be a loving person of sublime spiritual integrity who is prepared to suffer in God's cause. To follow God closely, which is necessary for the healing of the world, will not often bring popularity and acclaim, but more likely misunderstanding, rejection, and suffering. It is not a matter of going to look for trouble or deliberately provoking persecution. God does not call us to self-inflicted suffering. No! It is more a matter of being faithful to God in a world that has a very different agenda, and when faithfulness conflicts with conformity, there is often a cross to carry.

Isaiah 50 reads, "I gave my back to the bruisers, and my cheeks to those who ripped out the beard. I did not turn my face away from public humiliation and spitting."

God's servant is not inviting persecution but rather placing his suffering in the hands of God and not returning evil for evil. Isaiah is convinced that such is the redemptive thing to do. We may not quickly see the fruits of suffering love, but God will use such sacrifice for the ultimate salvation of the world. God will vindicate the suffering servant: "For the Lord my God helps me, that is why I am not in despair. I have set my face like a flint and I know I shall not be ashamed for my vindicator is near." (Isaiah 50:7)

In the economy of God, suffering for the sake of righteousness can be redemptive. From one angle, that is not a message I like to hear. I would prefer to believe that goodness would naturally prosper and bear fruit without pain, and that the world would witness the works of righteousness, applaud, and follow suit. I would prefer to think that loving servants of God would, after a long and successful life, die peacefully in their beds and have all people speak well of them at their funerals. I do not want to hear that goodness is often rewarded with misunderstanding and rejection.

Some people who have gone through severe persecutions say that it is not what happens to you that matters, but how you deal with what happens. Suffering can disillusion us, embitter us, and break us—or we can let God use it for a greater purpose and, in the process, become ennobled by it.

Chapter Seven

The Healing Power of Unlimited Love in Mind and Body, and in Eliminating the Arrogant Conflict between Religions That Gravely Threatens the Human Future

H EALING MEANT A LOT to Sir John, and he was enthusiastic about studies on healing in relationship to prayer and the experience of Unlimited Love. Equally important, he believed that Unlimited Love in the great religious traditions is our best hope for a healing human future not marred by incessant hatred and violence born of arrogant, insular, in-group love that demonizes people who do not believe in the same details that we do. So in this chapter I have chosen to combine these two key spects of healing as Sir John valued them both so highly. Unlimited Love was his answer to the arrogance of religious conflict. The human problem is a religious problem, and peace will only come, argued Sir John, when we abide in Unlimited Love and it in all of us.

LOVE HEALS

Sir John's true emphasis was that "love heals" (2000b, 16). "Constant, unconditional love," he writes, "can communicate itself to even the most badly abused among us. Love is one power that can eventually cut through the obstacles" (2012, 164). And "Love's energy is a healing balm." (2012, 159)

Love does heal. The compassionate clinician who loves his or her patients by attentive listening, carefully selected words of affirmation, a facial expression and tone of empathy, and continuity of care has

performed a medical intervention as important as any other. The typical patient is more able to adhere to difficult treatments and follow-up,
more likely to divulge useful diagnostic information, and less stressed
as a result. Unlimited Love in the form of a white light transformed Bill
Wilson during detox and he never took another drink as he went on in
love to cofound Alcoholics Anonymous, a mutual aid program in which
millions of alcoholics have been healed by spirituality, moral self-assessment, and helping others (the twelfth step). Every good clinician knows
that love contributes to healing.

Sir John, of course, availed himself of every technical advance in contemporary medicine and surgery, and he celebrated medical advances.
Like all of us, he wanted a doctor who was technically competent and
also dependable when it comes to tender loving care. Both his son
and his daughter were awarded medical degrees, one (Dr. John M.
Templeton Jr., MD) from Harvard and the other (Dr. Anne Templeton Zimmerman, MD) from Case Western Reserve University. Anne
Templeton Zimmerman's outstanding daughter Renee Zimmerman
also was awarded the MD degree from Case Western when I was on
the faculty there, and she was a distinguished student leader. I make
these observations to dispel any misconceptions; Sir John was a man of
modern medicine who celebrated its technological advances. Yet he also
understood that "love heals," and it always has.

One morning before a meeting in Nassau at the Paradise Island Hotel
I came down to the dining room early at about seven o'clock. Sir John
was sitting there alone in the empty room in a light green blazer, and
he motioned me over. He said, "Stephen, never forget that love heals
mental illness." "Well, Sir John, I know it can," was my response. And we
started talking about depression, personality disorders, and even schizo-
phrenia. Toward the end, we talked a lot about the different forms of
dementia. Sir John understood himself to a significant degree as a healer
of people's minds and bodies in the sense that if only they would abide
in the core principles of human virtue, control their minds, and pursue
education with eagerness, they would likely live happier, healthier, and
possibly longer lives. A chapter title in one of his books is, "Does Love

Act as a Healing Agent in Every Area of Life?" (2000b, 44). We must note the phrase, "in every area of life." It is not just in the clinic that love heals, and it heals in ways that go beyond illness to the level of family, society, and world peace.

Sir John never neglected to remind readers that love itself is vital to healing, as important as most technological interventions. So why wouldn't the experience of Unlimited Love be in various respects healing? Hence—his interest in spirituality and health. Unlimited Love frees people and inspires them, but does it heal them emotionally as well, and then, within certain limits, physically? It seems plausible that Unlimited Love would be healing. Imagine feeling a love astonishingly more affirming than we ever experienced even from our mother or father or closest friend. The agent of this love seeks nothing from us, is not pursuing personal gratification, and is concerned only with our complete well-being and flourishing. We are loved just as we are, just for existing. This love did not have to be earned, but it is there for us despite all of our shortcomings, imperfections, and disappointments. We feel dissolved into this Unlimited Love and at one with all humanity and with the God from whom this Unlimited Love emanates. A set of characteristic "spiritual emotions" ensues. Mere interpersonal gratitude is transposed into the higher key of prayerful thanksgiving before the Creator; mere optimism is transposed into a profound and enduring hope in God's agency and will; mere happiness is transposed into a deeper joy that is inwardly deep and not diminished by external circumstances. Thanksgiving, hope, and joy are interwoven with an inner freedom from self and an associated feeling of a concern for others that goes beyond the normal human repertoire with regard to depth of love. This sounds healthier than the opposites.

So let us draw a definition of healing care or love from the perspective of the giver: When the happiness and security of another means a great deal to me, I love that person. This definition holds in all cases. We may be looking over the crib of a young child, offering attentive listening to a friend, or helping a spouse. In a less concentrated and focused way, we can find deep meaning in the happiness and security of all human beings, from the acquaintances down the block who are grieving a loss to all the sick people in the local hospital. Some people very clearly feel intense love for some specific needful group, such as the homeless or the dying, and they devote their lives in active service, often inspiring others to join the cause. I can understand that some people wish to reserve "love" just for those few insular relationships with their nearest and dearest, but the word is used here in its more deeply traditional sense as a way of relating to all others in a warm and generous manner. Love has to do with one's heartfelt affirmation of the people one encounters in the course of the day, as contrasted with manipulation, hostility, and indifference.

A person who is severely ill often suffers an exquisite anxiety because all of the hopes and relationships held dear are imperiled. Routines are rudely interrupted, like a tsunami flooding the house and sweeping away everything and everyone in its path. Life is made much harder by small uncaring actions or words until finding the right caring health professionals, which usually makes all the difference in the ability to navigate through treatments that can be very challenging.

Imagine being ill or having a major accident that leaves you in a wheelchair for life. Some really caring physician, nurse, social worker, or pastor is able to connect with you, such that you feel that your happiness and security mean a great deal to that person, perhaps as much as their own or even more. This feeling is a bit like floating in a warm pool. You feel free for that short while from anxiety, rumination, fear, anger, bitterness, despair, and all other negative emotions that come with the territory of illness and that perhaps just yesterday were fueled by a callous and demeaning statement. But now you are grateful for being alive in this world, feel shielded and protected from stress, and are able to gather the strength to move ahead. You feel at rest and in fact may even be able to laugh, which is indeed healthy. This is all part of the affirmation, tranquility, and security of being loved. Someone set herself aside and bestowed her full and attentive presence.

This person did not do anything "big" at all. She did not put herself at any risk, although sometimes risks do come with the territory of being a health-care professional. She simply behaved with warmth, kindness, patience, and understanding. Perhaps she just asked you, in a caring tone of voice and with an empathic expression, "Is there anything you might want to make your stay with us a little more comfortable?" Small is beautiful. The quality of your experience as a patient is mostly the accumulation of such small interactions that leave you feeling respected and cared about. The kind professional or valet or custodian probably feels like a million dollars just for having been able to make your stay a little easier. We emphasize small acts done with care. These acts heal, and they are themselves a form of medical intervention and treatment.

Sir John wondered, "Do loving and feeling loved provide a sense of order and offer reassurances of more positive outcomes? Can believing that we are loved, especially by the Creator, create a sense of optimism and hope that may be physiologically beneficial? How might scientific research confirm this possibility?" (2003, 92). He asked, "How do giving and receiving love enhance our capabilities of promoting health, preventing disease, prolonging life, and hastening recovery?" (2003, 92)

In chapter 13 of *The Essential Worldwide Laws of Life* (2012), titled "Sharing Love," Sir John writes that love is like the sun, for it sustains itself. "The energy of love flows within us, changing and enlarging us," he writes. It replaces bitterness with acceptance and joy. "Love's energy is a healing balm. When we allow love to express itself though us as our basic nature, it automatically radiates out to every aspect of our lives," he notes.

He asks, "If we focused our thoughts on health and wholeness and immersed our emotions in unlimited love, could healing be a natural result? Someone described the healing energy of love as a spiritual antibody that helps eliminate disease in the human body. As we learn to be more loving and forgiving, unlimited love and forgiveness can influence our mental, emotional, and physical health." (2003, 93)

More broadly, Sir John sees love as the key to social health: "Scholars throughout the ages have defined love as the power that joins and binds the universe and everything in it; and love is often called the greatest harmonizing principle known to man. Teilhard de Chardin observed that love is the '*only force that can make things one without destroying them.*'

Love has the power to transform lives, to heal sickness, to mute evil, and to create harmony out of discord" (2000b, 7). He continued later,

> Love is an inherent power that, if allowed to be expressed in one's life, can transform disharmony, heal disease, and transmute negative conditions into part of a harmonious whole. The results of love are always good. But do not confuse sentiment and sympathy with love. Our focus in this book is the purified, transcendent power of divine love that is expressed through our hearts and minds when we are open and receptive to it and when we recognize, understand, and encourage love in its unlimited capacity. (2000b, 9–10)

Sir John often encouraged nonviolent solutions to world conflict, whenever plausible. He clearly recommended ultimate forgiveness of enemies, even if they must be forcefully deterred. As he would write, at least we must pray for their souls. Sir John concludes that we ought not to live "an eye for an eye and a tooth for a tooth," for "the world would soon be blind and toothless. Agape points to a different way—the way of Jesus, Gandhi, Buddha, and many others who were prepared to die, or to allow their selfish desires to die, but were not prepared to abandon agape." (1999, 99)

HEALING RELIGIOUS AND THEREFORE GLOBAL CONFLICTS

Sir John hoped that progress in spirituality along the awareness articulated in the above chapters would give rise to a future of peace, humility, and freedom from the arrogance and destruction of religious conflicts. Sir John wrote,

> In humility we have an opportunity to learn from one another, for it enables us to be open to each other and see things from the other person's point of view. We may also share our views with the other person freely. It is by humility that we avoid

the sins of pride and intolerance and avoid religious strife. Muhammad, the Prophet of Islam, stated, "Whoever has in his heart even so much as a rice grain of pride cannot enter into paradise." Humility opens the door to the realms of the spirit, and to research and progress in religion. Humility is the gateway to knowledge. (2012, 135)

Thus, a world of peace must be a world of humility, particularly if we are to ameliorate religious strife.

However sociopathic some people's conceptions of God have been, however deplorable the deeds done in the name of that God, and however much neo-atheists harp on this dismal history of arrogance and ignorance, such conceptions have absolutely no relevance to the question of the existence of God or the validity of spirituality. Sociopathic images of God are merely human projections made in ignorance and arrogance. It hurts to be badly mischaracterized and then to see people act destructively in your name, and I believe that John Templeton assumed that God feels this way. It might be argued that somehow all religion gets caught up in sociopathic images, and they are forever embroiled in territorial in-group–out-group conflicts. In fact this condition is far from true and certainly not inevitable. Clearly humanity must make "progress in religion," as Sir John emphasized, which will come mainly through the enhanced teaching, understanding, and practice of Unlimited Love. For Sir John, as for most thoughtful spiritual and religious individuals, this direction is the one in which to head.

This common direction never implied for him that different religions would lose their distinct identities, but rather that they would discover more deeply their true identities and actually compete to abide more deeply in Unlimited Love. Sir John did not wish to even use the word "religion" to describe any organization that does not teach and practice love.

Christian Inclusivism

Again, why the language of Unlimited Love? Sir John answered this rather simply:

While the expression "Unlimited Love" resonates closely with the ideal of *agape* love, it seems to be more appealing to many people than a technical, theological term from ancient Greece. Also, unlimited love is free of association with any one religious faith tradition and can appeal across cultures and academic disciplines. Unlimited Love can be representative of the ultimate nature of love as a spiritual and creative influence. (2003, 88)

In short, we live in a world where not too many people speak Greek, and where people approach love with different spiritual-religious worldviews. While Sir John did nevertheless continue to use the word *agape*, he mostly shifted over to *Unlimited Love*.

What about Christianity and other world religions? People ask where Sir John stood with regard to the value of the myriad of spiritual and religious traditions. Here the worship of a God of love is the key. He valued all spiritual and religious traditions to whatever extent they include within them insights into this underlying Unlimited Love that lies at the heart of the life of Christ. When Sir John Templeton wrote of "progress in religion," he meant in significant degree progress in Unlimited Love:

The rich variety of world religions creates a tapestry of amazing beauty—a testimony to the essential spiritual nature of our human existence. And yet, within this amazing and sometimes fascinating diversity can be found an equally amazing unity, the basis of which is "love." Perhaps without even being fully aware of it, religious leaders and their followers through the ages have defined religion largely in terms of love. All the world's great religions, to varying degrees, both teach and assume the priority of love in religious practice.

To put it another way, whether consciously or subconsciously, the world seems to have determined that any system of beliefs that teaches or tolerates hatred or even apathy toward others does not deserve to be considered a religion in the first place. (1999, 2)

Clearly Sir John was not a simple-minded religious pluralist, because this would entail endorsing spiritual and religious expressions entirely divorced from Unlimited Love. There is, after all, manifest harm and hatred in the domain of religions, just as there is in any other human domain, from business and politics to marriage and family. Like every thinking person, Sir John had a healthy ambivalence about religions in that they can bring out the very best and the very worst in human nature. Thus, he shared with Tillich a Christian inclusivism that can be distinguished from exclusivism (the idea that God cannot be active in religions other than Christianity), and pluralism (the idea that every religion or spirituality is equally loving and beneficial). He always encouraged us to learn whatever we could from all traditions that teach and aspire to greater love. Here Sir John was in accord with Fillmore and Unity, which while loyal to the Christian message were open to the work of God in various traditions consistent with universal spiritual principles.

Rescuing God from Sociopathic Images

Sir John Templeton was optimistic that humanity would eventually rescue God entirely from the clutches of arrogance and "small ideas." For every example of rage, fragmentation, and violence between the three Abrahamic faith traditions, we can take great hope in numerous examples of mutual appreciation, acceptance, and harmony. The solution Sir John espoused was to encourage these traditions to adhere more consistently and intensely to the teachings of a God of Unlimited Love that do in fact exist in each tradition (Levin and Post 2010).

Sir John believed that human ego has been the cause of religious wars over differences in creeds. He saw so much doctrine in religions as transient and changing over the centuries, such that "what is sometimes called heresy at one time is accepted as orthodox and infallible doctrine in another age" (2000a, 33). What really matters is the spiritual life, which is essentially "love to God and love to man" (quoting Theodore Parker; 2000a, 33). It is, he writes, arrogance that infects religions.

If this sounds a little unrealistic, the approach is still more realistic than the major alternative: the secular assertion that people in these traditions can or should take off their particular religious identities like clothes removed before a shower. Some people, like the wonderful bard of love John Lennon, do suggest a world "with no religion too," but most modern people around the world, including highly educated ones, define themselves—their core identity, their values, their ultimate commitments—in terms of their faith traditions. Human beings by nature ask cosmic questions about purpose and the why of the universe. The will to believe is as strong as ever. Thus, the more realistic approach is to highlight, develop, and make progress in Unlimited Love as it is articulated in the depth and particularity of each faith tradition. While every tradition has its distorting elements, most arose in the historical context of competition with other faiths. Sir John felt that the future could be brightened by focusing our spiritual minds on the love-based elements of these traditions, making humble progress in religion, and opening a mutually respectful form of competition in love. He understood that one response to religious differences and "wars between religions" is "to ignore all religion" (2000a, 29). But Sir John distinguished religion at its best from "the great havoc and suffering caused not by religion but by people who thought their own concept of God was the only one worthy of belief" (2000a, 29). The atheists, he asserted, are only rejecting some small and limited picture of God, and he was correct (2000a, 29).

Sir John wrote, "A peace-generating possibility at the heart of fruitful religion is the willingness to seek truths in other religions" (2000a, 30). In a quote that I have admired, he wrote,

> Differing concepts of divinity have developed in different cultures. Should anyone say that God can be reached by only one path? Such exclusiveness lacks humility because it presumes that a human can comprehend the fullness of truth. The humble person is ready to admit and welcome the various manifestations of unlimited divinity. (2000a, 30)

At their best, each tradition has indeed deepened the spirit of peace in the world—from Damascus where Jews, Muslims, and Christians have lived in friendship for centuries, to those in each tradition who have rescued those in other traditions from abuse, violence, and genocide. Every tradition has its high exemplars of Unlimited Love, even as they meet opposition from those who only love people who believe exactly as they believe.

Sir John was a Christian inclusivist. In this sense he greatly appreciated religious traditions on the condition that they contribute to lives of greater love. With this caveat in place, Sir John could be very open and appreciative of a wide variety of traditions:

> To seek to persuade all people to believe in one perspective would be a great tragedy. The wide diversity of faiths and theologies is a precious aspect of the richness of religion despite the fact that from a scientific point of view it can seem flawed and relativized. What is needed may be creative interaction and competition in a spirit of mutuality, respect and shared exploration. Were a drab uniformity to be accepted, the possibility of spiritual progress would be diminished. Learning to live with and learn from a rich multiplicity of spiritual perspectives is a step forward, because the more we know, the more we know we do not know. (2000a, 18)

He hoped for creative interaction, competition, and progress among religious traditions.

Toward the end of his life, Sir John expressed deepening concern over these faith traditions because of continued conflict that he counted among the most destructive impediments to human progress. While he acknowledged the importance of mutual acceptance between the Abrahamic faiths, he saw tolerance as a rather minimalistic rung in a ladder of attitudinal progress—rising upward to respect, trust, and ultimately to an authentic love between three faith traditions that share a common vision of a God of Unlimited Love. The theme of love between the adherents of the three Abrahamic faith traditions shaped Sir John's development of humility theology. Concerned with the potential for terrorism, he wrote *Pure Unlimited Love: An Eternal Creative Force and Blessing Taught by All Religions* (2000). Sir John cited passages from the scriptures and sages of the three faiths and from non-Abrahamic traditions underscoring God's love for all people without exception. He emphasized "how little we know" about spiritual realities, and how much progress could be made if these religions would all aspire to take to heart and learn more about the one great aspiration that they hold in common—Unlimited Love.

Each of these faith traditions offers guidance on how we may grow toward identifying with a shared humanity rather than a mere fragment thereof, and on how each of us may come to see ourselves in the other; each tradition at its best seeks to shape a normative religious experience that guides its adherents toward recognition of a common humanity with believers on different spiritual paths. However, each tradition also contains restrictive elements that focus in a purely insular direction that may even devalue or demonize nonadherents. The Abrahamic traditions are not identical in this regard, yet each can enunciate its commitment to Unlimited Love more vividly, and can practice it more extensively with dedicated minds and hearts.

Extensivity

Due to an emphasis on tribalistic elements in fundamentalist religion, over and above Unlimited Love, norms of religious devotion may be observed to elicit harmful tendencies. The glorification of a favored identity (e.g., election, salvation) has made fundamentalism vulnerable to reifying the "otherness" of outsiders. This stance distorts the messages of love, unity, justice, compassion, kindness, and mercy that lie at the core of the great orthodox traditions, serve to instill humility in religious believers, and are consistent with Divine Love.

In-group identity is, of course, a part of the human need for security and community. Tribalism at its best serves to reaffirm and elevate the holiness of a people and thus ideally inspire and enable greater and more successful acts of communal and worldly service to exemplify that holiness. What must be avoided is what the theologian H. Richard

Niebuhr referred to as "henotheism," the deification of the in-group and the correlative demonization of outsiders. This is the height of religious arrogance. The enemy of progress is not tribalism per se but rather the exploitation of tribalistic impulses dormant in respective religious traditions in order to marginalize, condemn, and attack other tribes, rather than to achieve Unlimited Love.

In his classic *The Ways and Power of Love* (1954/2002), Sorokin developed a measure of love that involves five aspects. The second aspect of love is *extensivity*: "The extensivity of love ranges from the zero point of love of oneself only, up to the love of all mankind, all living creatures, and the whole universe. Between the minimal and maximal degrees lies a vast scale of extensivities: love of one's own family, or a few friends, or love of the groups one belongs to—one's own clan, tribe, nationality, nation, religious, occupational, political, and other groups and associations" (1954/2002, 16). Sorokin had immense respect for family love and friendships, but he clearly thought that people of great love lean outward toward all humanity without exception, and that truly great lovers inspire others to do the same. He understood human beings to have pronounced tendencies toward insular group love, and he argued that religion at its best moves agents beyond their insularities to humanity and even all life.

Sorokin was a scientific optimist, hoping that enhanced understanding might unlock the "enormous power of creative love" (1954/2002, 48) to stop aggression and enmity and contribute to vitality and longevity (1954/2002, 60–61), cure mental illness, sustain creativity in the individual and in social movements, and provide the only sure foundation for ethical life. Sorokin's general law is as follows:

> *If unselfish love does not extend over the whole of mankind, if it is confined within one group—a given family, tribe, nation, race, religious denomination, political party, trade union, caste, social class or any part of humanity—such in-group altruism tends to generate an out-group antagonism.* And the more intense and exclusive the in-group solidarity of its members, the more

unavoidable are the clashes between the group and the rest of humanity. (1954/2002, 459, italics in original)

In-group exclusivism has "killed more human beings and destroyed more cities and villages than all the epidemics, hurricanes, storms, floods, earthquakes, and volcanic eruptions taken together. It has brought upon mankind more suffering than any other catastrophe" (1954/2002, 461). What is needed, argues Sorokin, is enhanced extensivity (Levin and Post 2010). Sir John would fully agree.

Sorokin placed his faith in science, as did Sir John:

Science can render an inestimable service to this task by inventory of the known and invention of the new effective techniques of altruistic ennoblement of individuals, social institutions, and culture. Our enormous ignorance of love's properties, of the efficient ways of its production, accumulation, and distribution, of the efficacious ways of moral transformation has been stressed many times in this work. (Sorokin 1954/2002, 477)

Sir John Templeton understood that world peace will only come when we find peace between religions, for these are essential aspects of world cultures and civilizations. Especially in the late 1990s, he began to write about Unlimited Love across all worthy traditions, and about honoring this ideal in every tradition so that the humble best in human nature rather than the arrogant worst might be unveiled.

Sir John wrote *Agape Love: A Tradition Found in Eight World Religions* (1999) as a call for all religions to rededicate themselves to the overarching law of Unlimited Love:

The rich variety of world religions creates a tapestry of amazing beauty—a testimony to the essential spiritual nature of our human existence. And yet, within this amazing, and sometimes fascinating, diversity can be found an equally amazing unity, the basis of which is "love." (1999, 2)

Sir John was greatly concerned with intolerance. In a very powerful passage he wrote about the role that diligence of mind and humility can play in moving us forward:

> It can be a religious virtue reverently to cherish scriptural beliefs and to study them with the utmost seriousness. But of course a reverse side of this virtue can be a vice of intolerance. Is it easy to become intolerant if we are not diligent to guard our minds actively to be humble and to remember that despite differences in religious traditions we all have profoundly limited concepts with respect to the vast divine realities? *Can love and the vastness of divinity reduce our differences as we seek to understand by a variety of different ways and through many various traditions?* Can diligence and humility help heal conflict between many communities holding different religious points of view? (2000a, 28, italics added)

Here again Sir John finds redemptive hope in the self-control of the mind and a diligent discipline of thought, coupled with the virtue of humility.

Sir John was not content with the minimalist language of tolerance, and he distinguished it from the humble approach as follows:

> However, tolerance is not the same as the humble approach. Can we seek to benefit from the inspiring highlights of other denominations and religions, not just to tolerate them? Should we try our very best to give beauties of our religions to others, because sharing our most prized possessions enhances the highlights of each? (2000a, 31)

COMPETITION IN UNLIMITED LOVE
AS ALTERNATIVE TO RELIGIOUS CONFLICT

Sir John loved competitions! He loved them in every domain of life, including investing, essay contests, and even in Unlimited Love! For religions to handle their differences, he argued, we need free and open competition in love. May the religions with the most enhancing and positive ideas win: "To best deal with conflicts over dogmatism, possibly we can benefit most by listening carefully and respectfully rather than arguing. If a message is not truth, it will fade away in time and especially if made to compete in the open air with other more broad-minded ideas" (2000a, 36). He championed the use of religious tolerance to promote "constructive competition" that allows each faith tradition to demonstrate its value and goodness. (2000a, 123)

Even with regard to new religious movements, Sir John had a tone of calm appreciation:

> Should we be gentle, kind and sympathetic toward new prophets even though they bring new ideas strange to us? Should laws forbid religious expression and research, however misguided we think them to be? Was any useful purpose served when the Inquisition forced Galileo to recant in 1633? Jesus, too, was considered unorthodox by the learned religious establishment of his day. Beneficial originators in other great religions also were often called heretics. Abraham and Moses were considered heretics by neighboring tribes in their old age. (2000a, 38)

As creeds wax and wane, love lives on and on.

We should thus not react to new ideas but let them survive based on their value in competition with other ideas. As Sir John concludes what I consider to be the most powerful chapter I have ever read on the root causes of religious conflict and the potential solutions in humility, he writes, "There is room for many branches on the tree. The life sap of pure unlimited love lives on and on" (2000a, 39). People really in touch with this "life sap of pure unlimited love" were understood by Sir John to be creative spiritual geniuses. Most of them ("Buddha, Paul, Zoroaster, Muhammad, Wycliffe, Luther, Calvin, Wesley, Fox, Smith, Emerson, Bahaullah and Eddy," to cite one of Sir John's lists in 2000a,

42) were deemed heretics and persecuted. Sir John believed that progress occurs through such geniuses. "Why does God's process of evolution produce these rare geniuses on earth? Is it the divine plan that they should help all people to progress?" (2000a, 43). He cites Mother Teresa of Calcutta, Chiara Lubich, and others who have won the Templeton Prize for Progress in Religion as possible "humble co-creators with god." (2000a, 46)

Sir John objected to the idea that the U.S. Constitution forbids religion in the public schools:

> The founders of our nation intended to ensure free and fair competition among religions, not to stamp out religion altogether. Their efforts to separate church and state were not efforts to abolish religious education in the classroom. The current practice of near total omission of religion from schools prevents free competition. The net effect may be to imply that only secular knowledge is respectable. Are new generations being educated to be intellectual and cultural adults but spiritual and ethical infants?" (2000a, 124)

Furthermore,

> Are there not ways, however, to keep religion in our schools without favoring one religion or denomination over the others? Methods have been devised to allow each child, or its parents, to choose from a broad spectrum of religious studies—from traditional world religions to atheism. Would it not be possible, for example, for each school to provide a room which could be used as chosen hours for thirty minutes weekly or more by various religious or nonreligious groups wanting to present their beliefs or to engage in worship? Students or their parents could then choose which to attend for such thirty minutes of spiritual education weekly or more often. (2000a, 125)

Sir John knew well that in-group exclusivism has killed more human beings and destroyed more cities and villages than any other force in the universe. He hoped that science would provide an inestimable service by better understanding the ways and power of Unlimited Love and encouraging religious traditions to live up to it. For Sir John, the main obstacle to religions abiding in Unlimited Love is arrogance, especially with regard to theology: "In every major religion wars have been fought about differences of creeds. Nations and tribes have exterminated others because they worshipped different gods or the same god as taught by different prophets. This is human ego run wild. Let us humbly admit how very small is the measure of men's minds. This realization helps to prevent religious conflicts, and obviates attacks by atheists against religion." (1981/1995, 47)

Importantly, Sir John did not wish in the least to diminish the richness of religious pluralism and particularity. He hoped that every religion, in all its particularity, would be able to abide in Unlimited Love:

> The purpose of this book is not to conclude that all religions are the same, for certainly they are not. Nor is the goal to try to convert anyone from one religion to another. Rather, the purpose is to point toward the possibilities and respon- sibilities of love. It is to awaken people to the realization that despite differences, all religions share some very important, fundamental principles and goals, the highest of which is the realization of agape love—unconditional, unlimited, pure love. (1999, 5)

Thus Sir John was attempting to find the universal through the very particular windows of distinct traditions, each of which has its own beliefs, practices, and history. In other words, he was rejecting the idea that there is a "view from nowhere" into Unlimited Love. As he wrote, "When we embrace the possibility of agape love, we are expressing, amidst our differences, a unity of purpose, a common hope. At the dawn of the third millennium, what vision could be more important?"

Sir John wanted the whole world someday to realize the full potential of Unlimited Love. He saw love as the fundamental motivator of most of what we do, influencing everything from family dynamics and health to inner peace and global politics. Evolution, religions, neuroscience, social science, biomedicine, spirituality, and human development increasingly converge around the power of love in our lives, as we leave our islands of emotional isolation and do things for others. When our financial security is shaken, when our health is compromised, and when disappointments are heavy, generous love remains our saving grace and dignity, the absolute sine qua non of a more meaningful, happier, and healthier life. We cannot control all the external circumstances in life, but we can determine to respond to those circumstances in ways that do not betray the spirit of love.

*Three Primary Evidences of Sir John's
Thesis Consistent with His Core Ideas
and Statements Focusing on Their
Current Scientific Plausibility*

Introduction to Part 2

THREE EVIDENCES FOR SIR JOHN'S THESIS

I N 2001, when Sir John invited me to start the Institute for Research on Unlimited Love, he emphasized three areas of evidence for the reality of Unlimited Love. I agreed with him then and still do today.

First, do people self-report the spiritual experience of Unlimited Love widely, and how do they describe its effects on their lives and on their interactions with others? Sir John urged media outlets to allow people the chance to tell of their experiences. He wondered if feeling God's love occurs in many or most people's life experience, and how it might impact their levels of generosity and benevolence not just with regard to the near and dear but with regard to humanity equally considered. If levels of benevolence are significantly increased, this would indicate that such a self-reported spiritual experience kindles and enlivens love. Sir John had supported research on the topic of "spiritual transformation," a very broad category, but he wanted more focus on the specific experience of agape or Unlimited Love and how that experience changes people. It took nine years to assemble the right research team to conduct a random scientific survey of the American people. The results show that the self-reported experience of God's love is very widespread in the United States, and that by both self-report and peer assessment it does in fact expand and intensify benevolent actions. It would be useful to conduct a worldwide survey in the future.

Second, do people who sincerely love God and their neighbor as themselves flourish? Are they happier, healthier, resilient, and more purposeful? Are they blessed with abundance of years? Do saintly individuals live longer (after factoring out those who die young due to martyrdom)? Time and again in his writings, Sir John emphasized the centrality of the double love commandment, "Thou shalt love the Lord thy God with all thy heart, and with all thy soul, and with all thy mind," and "Thou shalt love thy neighbor as thyself" (Matthew 22:37–40). He asked me to do whatever I could to show that individuals who abide in this love of God, neighbor, and self in fact prosper. I refer to this thesis as the *ontological generality*, which means that we are shaped and formed as creatures to realize our fullest thriving and success in this tripartite communion of love. When we follow the Golden Rule of doing unto others, do we generally flourish? And do we flourish even more when we place this rule in the wider context of the double love commandment? In other words, it there something about doing unto others while in a state of feeling God's Unlimited Love that amplifies the benefits to the human agent? Based on a number of our investigations, this appears to be the case.

Third, Sir John wanted to keep a focus on physics and cosmology, from the big bang to quantum reality, because eventually the human experience of Unlimited Love needs to be connected to the objective universe "in which we live, and move, and have our being"—another of his favorite quotes from the writings of Paul. At one brief meeting in Nassau, Sir John said to me, "Stephen, Genesis says, 'In the beginning God.' And then the Gospel of John says, 'God is love.' So then in the beginning was Love, and this is the root of reality and the universe." Both Sir John's Presbyterianism and his engagement with the metaphysical ideas of the Unity school of Christianity contributed to his hope that science might at least raise the big question of the objective reality of the Ground of Being or Ultimate Reality (aka "God") as Unlimited Love. He understood that it might take a hundred years for science to explore this question, and he had no idea if in the end science could answer it fully. But he was interested in the adventure. Sir John never funded a physics experiment, but he wrote extensively, humbly, and speculatively about where his metaphysics and physics might intersect, and he certainly believed himself that the idea of Unlimited Love as the prime creative energy underlying all that exists is at least scientifically plausible, however little we know to date. This is evident in his *Is God the Only Reality?* (1994).

Let me clarify the significance of these three areas of investigation with the help of a truly big question: *If people who self-report the experience of Ultimate Reality as Love really do flourish (elevated benevolence and joy), and if it is objectively plausible that such an Ultimate Reality exists as an energy of creative love underlying everything that exists, then can we responsibly state that the metaphysical idea of Unlimited Love as Ultimate Reality is at least scientifically plausible?* Below I restate these three areas of investigation more clearly, and I pursue each of them in a separate chapter of part 2 of this book.

- That People Widely Self Report Spiritual Experiences of Unlimited Love and an Associated Enhancement of Benevolence
- That People Who Love God and Their Neighbors as Themselves Generally Live Healthier and More Joyful Lives
- The Plausibility of an Underlying Ground of Being or Ultimate Reality (aka "God") That Constantly Creates and Sustains the Universe

Sir John Templeton was deeply grateful toward a God of Pure Unlimited Love whose love is the source of profound spiritual growth and transformation in the individual. He was convinced that when we love God, neighbor, and self rightly, we can flourish because this is our true and original nature and being; he thus did all he could to abide in love and to model this for others. Sir John was the kind of man who could lie down on the brown leather couch in his office and pray and meditate, knowing that every single thing that exists in this universe is constantly sustained by God and is constituted at the most basic level of being by God's infinite prime energy and Mind, including each cell and atom in his body, and every star in the sky. As he wrote,

> God is infinitely great and also infinitely small. He is present in each of our inmost thoughts, each of our trillions of body cells, and each of the wave patterns in each cell. (1990/2006, 204)

Chapter Eight

Unlimited Love as Ultimate Reality
in Sir John's Writings

L ET US BEGIN WITH repeating two paragrpahs from the letter Sir John wrote to me dated August 3, 2001:

> I am pleased indeed by your extensive plans for research on human love. I will be especially pleased if you find ways to devote a major part, perhaps as much as one-third of the grant from the Templeton Foundation, toward research evidences for love over a million times larger than human love. To clarify why I expect vast benefits for research in love which does not originate entirely with humans, I will airmail to you in the next few days some quotations from articles I have written on that subject.
>
> Is it pitifully self-centered to assume, if unconsciously, that all love originates with humans who are one temporary species on a single planet? Are humans created by love rather than humans creating love? Are humans yet able to perceive only a small fraction of unlimited love, and thereby serve as agents for the growth of unlimited love? As you have quoted in your memorandum, it is stated in John 1 that "God is love and he who dwells in love dwells in God and God in him."

This is quite a letter to ponder, and I must confess that, in retrospect, I only now begin to fully appreciate its depth. Later in the letter, Sir John asks, "Could love be older than the Big Bang?" With this, we have the question that most interested his amazingly curious mind.

In a brief letter to me written two years later on October 30, 2003, Sir John reiterated his concerns:

> Do your Advisors think that it will be helpful for you to subsidize a weekly email letter for thousands of possible donors, each letter being on a different aspect of *unlimited love*, beyond the love which is a product of humanity? Probably, top quality articles can be found about how *unlimited love* may have been the creative force 15 billion years before humanity began, and how *unlimited love* may be the force behind (rather than the product of) the amazing acceleration of creativity.

It could not have been clearer that Sir John was pushing the institute to address the biggest question of them all, which is whether Unlimited Love is the prime creative energy from which the universe originates.

Sir John wrote a letter to me dated February 25, 2004, that I took to be his most definitive vision statement with regard to why he made such a major grant to the Institute for Research on Unlimited Love:

> Congratulations on your diligent and extensive work for your foundation to encourage scientific methods of discovering over one hundredfold more of aspects and benefits of unlimited love. Your progress report which my son discussed at our meeting yesterday seems to imply that your program is limited to the type of love called *Agape*. My intention when making a major grant to your program was that it would encourage the top one millionth (intellectually) of the world's people to become enthusiastic about science research on every form of love, not only all types of human love but also increasing evidences that love may be the driving force in all creativity even before the Big Bang.
>
> For example, evidence is increasing that creativity itself

is a form of love and so is magnetism and gravity. The New Testament says that "for God is love and he who dwells in love dwells in God and God in him." I hope that your program can be expanded to include a new encyclopedia about the benefits and aspects and varieties of love. Love had been the central theme of all major religions for thousands of years, but until the latest two centuries methods of science have not been applied for human discoveries of over one hundredfold more about love.

God bless you and your important program.

So again and again, Sir John urged me to think about love as "the driving force in all creativity even before the Big Bang." We actually did produce an *Encyclopedia of Love in the World Religions* (ed. Yudit Kornberg Greenberg, 2 vols., 2008).

Regrettably, it took me a few years to catch up to Sir John intellectually. He was, I believe, a tad disappointed in the human-centeredness of much of what the institute was asking. I had not fully appreciated the metaphysical dimensionality of his approach, or perhaps I understood it but struggled to find ways to actually investigate it. No doubt in writing this book five years after Sir John's passing I am at level feeling the need now to proceed with the sort of metaphysical courage that Sir John was trying to teach me. He succeeded, but there was for me a learning curve. Love preceding the big bang, Ultimate Reality as a Love Creativity underlying all that is? This was, for Sir John, the biggest question area, and one where he was as bold as could be. He hypothesized an Infinite Intelligence of Creative Love that preexists all else—and that itself came from nowhere, had no beginning, had no source. It is that about which we cannot ask from whence it comes. It is ultimate, the most basic, the true origin. It comes from nowhere and nothing precedes it.

For Sir John the fingerprints of Ultimate Reality might be found not just from looking up and out at the universe with its sun, moon, stars, and infinity. He wondered about the ways in which Ultimate Reality might be present in every tiny particle of quantum physics as these have become better understood. Those quarks that we know exist but have not seen—these charged building blocks of the neutrons and protons that constitute the nucleus of the atom, these almost infinitesimally small particles that are tied together by the "strong force" that if unleashed can destroy cities—is all this energy and matter and force the handiwork of an awesome prime energy of Unlimited Love?

Sir John and Robert L. Herrmann saw God as a loving creative energy underlying and sustaining all that is in every moment:

> A wonderful statement about God is given in the first chapter of John. In the majestic language of the King James Version it reads, "In the beginning was the Word and the Word was God. All things were made though Him, and without Him was not anything made that was made. In Him was life, and the life was the light of men." This might be paraphrased in the language of modern science as, "God is creating the universe of time, and space, and of humankind. Creation proceeds from idea to word to sense data. The Creator is not completely visible to us but is the Universal Spirit, the causative Idea which sustains and dwells in all He created and is creating. The orderliness and lawfulness of nature and of spirit exhibit God to humankind." (1989/1998, 32)

All of reality is the boundless manifestation of the Word, of God's eternal creative energy of love and Divine Mind. Could the big bang be the creative act of Unlimited Love? Sir John wrote, "Ultimate reality can be vastly more astonishing than the sum of phenomena already observed. More and more, the immensity and complexity of the physical universe points to a nonphysical creator who is infinite" (2000a, 60).

AN ORIGINAL PRIME ENERGY OF UNLIMITED LOVE

Science changed its view of reality in the twentieth century. Things are more mysterious and profound than the old simple-minded materialist and mechanical view of the world. Sir John wrote, with Herrmann, as follows:

> Just a century ago science appeared to be tidying up our world, dispelling the illusions of gods and inexplicable miracles and finally providing us with an "objective" world. Yet now, despite the fact that we live at a time of enormous advances in our knowledge of our universe, there are few who would claim that objectivity is realizable in science, and fewer still who would exclude the possibility that an invisible hand is involved in this vast process. For our science in this century has brought us relativity and quantum theory, and with them have come some strange new understandings of our world; space and time are no longer separable, the passage of time is variable, dependent on the velocity of the mover, matter and energy are interchangeable, and cause and effect have no meaning at the level of sub-atomic interactions. The credo of objectivity, together with its tight little mechanisms and clockwork images, is gone. Matter has lost its tangibility and quantum physics has shown our world to be more like a symphony of wave forms in dynamic flux than some sort of mechanical contrivance. (1989/1998, 143)

Sir John is saying that reality is seemingly fluid and so much more complex than we have imagined. That it arises from a prime energy of Unlimited Love seems a plausible way of suggesting that whatever this underlying energy is, it has at its essence a creativity motivated in love.

Sir John and Herrmann conclude, "All that we have said seems to shout for a God or marvelous creativity who mediates a world in dynamic flux, a world moving in the direction of ever more complex, more highly integrated form—form set free to dance" (1989/1998, 162). Everything is unfixed and fluid, in stages of becoming and ceaseless flux. Thus, what is the Ground of Being? What underlies this complex reality? Sir John and Herrmann ask, "Perhaps in the end the only reality is God." (1989/1998, 163)

In a passage that captures Sir John's ideas with clarity, he and Herrmann wrote that subatomic physics indicates that matter is not what we thought it was several decades ago. We have the much deeper and more complex reality of "virtual particles boiling up out of nothing," implying for many the idea of a Sustainer who lies behind and underneath all things and "in whom we live and move and have our being" (1989/1998, 37). "Physical reality seems less and less tangible. Perhaps in the end it may only serve to point us to another awe-inspiring transcendent and immanent Reality." (1989/1998, 37)

This idea of God, this Ultimate Reality of Unlimited Love, gives us joy, inner peace, healing, and purpose, and sustains every aspect of the matter and energy that constitute the universe and our very selves. Such a God is not far away. This Ultimate Reality is *immanent* or manifested in and encompassing the material world, in contrast to theories of transcendence, where God is seen to be outside the material world rather than permeating it. For Sir John, God permeates everything, so our spiritual task is to grow in awareness of this and to recognize that we participate in God and God in us.

Ultimate Reality? Most readers will ask, what kind of reality are we talking about here? An answer comes from the perspective of theology and from that of physics. Sir John hoped that someday, perhaps a century into the future, the scientific methods of physics will develop to a level where the question of Ultimate Reality can be answered. The big question is, what is the ultimate nature of reality? To put it another way, the hypothesis could be that Ultimate Reality is a form of infinitely creative and loving pre-energy that exists before space and time; that preceded the big bang, as well as the creation of light, energy, mass, and the laws of the universe that allow for its stability; and that constitutes the underlying sustaining essence and matrix of everything that is. In terms of theology, this Ultimate Reality is the God who existed before creation and who said, "Let there be light."

Sir John and Herrmann wrote, "What we see gradually coming into focus will be a surprise to many. We conclude that the only coherent explanation, the grand organizing paradigm, to which both science and religion point but which neither begins to exhaust, is the God of the universe—what Loren Eiseley called 'the Great Face Behind'" (1989/1998, 8). Moreover, "The manifold scientific discoveries of the late twentieth century cause the visible and tangible to appear less and less real and point to a greater reality in the ongoing and accelerating creative process within the enormity of the vast unseen" (1989/1998, 9). A revealing passage from this work reads, "Perhaps, in the end, Reality will appear far deeper and more profound than our limited human abilities can ever hope to comprehend." (1989/1998, 25)

When the great spiritual and mystical thinkers ponder the experience of God, they often speak of an eternal being of light—bright, gentle, filled with love, and awesomely powerful. They speak of being absorbed in this light of love and of being transformed in a way that brings a special sense of joy and makes possible a deeper love of neighbor and of nature. This absorption in love and joy is almost childlike—a feeling that God's sole desire in creation is to see me joyful in the exuberance and peace of God's love. These mystics speak of an awareness of another dimension of reality and a sense of eternity beyond time and space in which they encounter the Supreme, the Source of all that is, the Absolute, the Ultimate Reality ever pure. They speak of shedding their egos and social masks in a realization of who they really are. They speak of the vastness of an ocean of love and truth, and in humility and awe before the One they appreciate the need to learn and to grow in love and wisdom. They speak of an eternal identity, of a consciousness of a nonmaterial soul that is like a point of light energized by the source light of Divine Love. They engage in prayer and meditative techniques that allow them to focus their minds, which include an eternal spiritual element linking them with the Divine Mind. They speak of a power of the individual mind that transcends the limits of time and distance in the physical universe. In this other dimension they feel at home, at rest, in an infinite expanse and presence of Pure Unlimited Love. This experience can become so absorbing that some of the mystics have very little to do with the active world of human interactions, but the truly great ones return to the world like fish who have jumped for a moment above the surface of the water into the shining sun and now have much to teach and do in the service of all mankind.

For those who understand God or Ultimate Reality in personal terms and for whom pure love is the essential motive in all that God does, then the question of why the big bang took place is easy to answer as a matter of faith: God created out of love for us to share in the experience of co-creatorship and Mind, to love God and be loved by God, to love neighbor as self, and thus to experience supreme joy. Is science ready to prove such a thing? No, absolutely not. The question is still too big for science to handle. Nothing is proven or disproven herein. But there may come a time when a complete theory of physics and the universe, a "theory of everything," can demonstrate this article of faith. Meanwhile, however, what physics can do is show the plausibility of the insight that an Unlimited Love could be the Ultimate Reality.

SIGNALS OF TRANSCENDENCE

Sir John with Robert L. Herrmann began *The God Who Would Be Known* as follows: "This is a book about signals of transcendence, about pointers to the Infinite that are coming to us not from mystics but instead through the recent findings of science" (1989/1998, 1). Here the co-authors take up the reverential attitude of great scientific minds of the past such as Francis Bacon, Isaac Newton, and their contemporaries. The very laws of the universe that Newton described reflected for him the constancy of God. But Newton's God, like that of Descartes, was at a distance and detached—a Creator who brought the universe into being but is not actually involved in it. Sir John and Herrmann prefer a divine activity that is "far more open-ended and immediate than the clockwork image would allow" (1989/1998, 21). The distant transcendent God is replaced by a God whose energy and love are literally "in" the world, in constant sustaining activity. All the laws and constants of the universe connect back to the Intelligence and Love of God, but God also upholds them in every moment. All the energy and material of the universe are expressions of God and Love and derived from God and part of God. God is manifest in the universe in every moment. The universe is contingent in that it depends entirely on God for its origin, continued existence, and its orderliness. The authors write,

> Admittedly, it was difficult in Newton's day to appreciate the need for the immanent Creator, constantly willing the order and constancy of his creation, especially as philosophers of that time attempted to define nature with all its multivariable phenomena in terms of exclusive mechanical law. But with the advent of Einstein and his theory of special relativity, a massive shift began in the way scientists viewed the physical world. (1989/1998, 129)

Relationships between space and time, matter and energy, and then quantum theory shook science to its foundations.

With these advances in science away from the idea of a material and mechanical universe to one that is much more fluid, malleable, and grounded in mysterious energies that we do not fully understand, it became plausible for Sir John and many spirituality-minded thinkers to posit the living creative presence of God in all that is, to find renewed purpose and joy in this presence, and to ask why the whole symphony of the universe as created, ordered, and constantly sustained would come into being for any motive less than Unlimited Love. Sir John considered himself part of a movement of sophisticated scientific thinkers who saw increasing rapprochement with a new theology of a continuously sustaining, ever-present, and imminent God of love. To quote again from *The God Who Would Be Known*,

> We began this book with the idea that the God who has made this awesome and wonderful universe is utterly beyond our capacity to measure and yet is also the God who would be known. He has placed remarkable signs in the heavens, on Earth, and in ourselves: signals of transcendence. In conclusion, much evidence indicates that this universe is here

by divine plan, and that science itself, for decades a bastion of unbelief, has once again become the source of humankind's assurance of intimate divine concern in our affairs. (1989/1998, 237)

At the conclusion of this book, Sir John and Herrmann speculate about future possibilities in science, focusing on the ideas of physicist David Bohm and neurophysiologist Karl Pribram—that the universe may be something like a "superhologram" in which "everything is connected to everything else by the immanent God, who holds the universe in being" (1989/1998, 241). Of course they qualify this comment by stating that this description of the universe as a "superhologram" may be a "gross speculation" in scientific terms (1989/1998, 241), but this passage shows how Sir John welcomed new ideas with an open mind and was convinced that "exploration of God's universe that is just beginning now becomes a new journey in spiritual discovery, a voyage into the sphere of the spirit" (1989/1998, 15). And the more we learn about God's universe through science, the more we are fulfilling our purpose in applying the Love-Gift of our minds in eagerness to learn, and the more we can be awakened into joy as we understand that the love and energy of God are right here with us in every moment in all things, awaiting only our coming into awareness.

Sir John saw in the big bang the idea that the entire universe was once concentrated into a single infinitesimal point, the plausibility of God's creative action. Everything that is springs forth from a blinding flash, and we find a striking resemblance to the passage from Genesis 1:3, "Let there be light." The speed of light, the masses of various particles, the equivalency of mass and energy ($E = mc^2$), all slowly reveal the Mind of the Maker.

The idea that Unlimited Love is Ultimate Reality seems to be unproven but not implausible. We know how much Sir John felt this "subjectively" in his effervescent and joyful celebration of creation along the beaches of Lyford Cay. The beauty of the natural world was for him a window into Divine Love. He also had a sense of awe or *mysterium tremendum* before the power and beauty of nature, and he included awe along with joy in those spiritual emotions that the Templeton Foundation should investigate. How "objective" does anyone need to get about this question of Ultimate Reality? Can't we just leave it to the mystics and poets? Sir John wanted to pursue the objective approach in the hope that eventually minds, which are God's Love-Gift to each of us, would be able to clarify and confirm that indeed Unlimited Love is Ultimate Reality. Presumably his appreciation for the emerging plausibility in mathematics and physics for an Infinite Intelligence prompted him to want to write a book titled *Unlimited Love as Ultimate Reality.*

So, is it possible that a Divine Mind of Unlimited Love thought into being all the laws and constants that underlie the universe and, through a big bang of creation, brought it all into being in such a way that, through an unguided evolutionary process of increasing complexity, intelligent creatures sharing in Divine Mind could eventually come into existence?

Some very solid minds in physics do not endorse some form of Infinite Intelligence. The British astronomer Sir Martin Rees, 2011 winner of the Templeton Prize, in his book *Just Six Numbers: The Deep Forces That Shape the Universe* (2000), shows that six numbers define the basis of material reality and of physical constants such as gravitation and electromagnetism. But Rees remains agnostic, leaving such questions in the domain of personal subjectivity. Another good example of such a mind is Leonard Susskind, who in *The Cosmic Landscape* (2006) argues that while it is easy to surmise that an Infinite Intelligence underlies the universe, in fact this astonishing order in our universe can be explained by "string theory," an area of physics with which Susskind is credited with much of the original theorizing.

No one is suggesting that anything has been proven. Sir John felt that this kind of proof, if possible, would be off in the future, and he did not make any assumptions other than that the best scientific methods operating in complete autonomy would lead to progress in knowledge about these very big questions.

But Sir John did have metaphysical leanings that suggested to him that Ultimate Reality is a form of Unlimited Love. Spiritual traditions all speak of a Being or Godhead that precedes creation, including time and space. We don't need to trace here the ideas of perennial philosophy, but traditions do converge closely—for the Hindus Brahman, for the Kabbalists *Ein-Sof,* for the Neoplatonists the *Deus Absconditus.* This Being is originally "in the beginning" unmanifest and hidden, Pure Mind or Spirit, and Pure Love and Creativity. But there is incompleteness in this Godhead, a creative propensity to move from being to activity through creation, perhaps because of the need to be known by a creature whose mind is actually a small portion of the Divine Mind itself. This Godhead is perfect love, or "Pure Unlimited Love," to use Sir John's language. All the energetic and material universe is in some way made up of this source of all that is, this Ground of all Being. In physics these days it is noncontroversial to state that particles arise out of pure energy. So the material and visible world is really a kind of phase change in which time and space come into existence with matter and energy from Infinite Mind? Sir John surmised that the world we live in is created and sustained in Divine Idea.

Contemporary physicist Paul Davies, another Templeton Prize winner, in his book *The Mind of God* writes thus:

> Through my scientific work I have come to believe more and more strongly that the physical universe is put together with an ingenuity so astonishing that I cannot accept it merely as brute fact. There must, it seems to me, be a deeper level of explanation. Furthermore, I have come to the point of view that mind—i.e., conscious awareness of the world—is not a meaningless and incidental quirk of nature, but an absolutely fundamental facet of reality. (1992, 16)

In other words, consciousness is not a side effect or by-product of matter, but rather the reverse is true.

Davies would not go so far, but for Sir John our minds are the Love-Gift of the Divine Mind and make us the image of God; more literally, we are one with God even if we are unaware of this. Shaping it all is a purposeful, all-loving God who wants to be known by and to know each one of us, and in this lies mutual love and consequent joy. This whole drama involves (1) an unseen Godhead, (2) God's manifestation in creation (the big bang, the laws of physics and math, and a sustaining underlying presence), and (3) the eventual evolution of a creature upon whom Mind is bestowed. Of course, this may never be proven strictly, but if a complete theory of physics ever does come to this conclusion, it will be resonant with a perennial human experience and philosophy that never seems to fade and is typically reinvigorated despite times of rampant materialism, sensate culture, and moral decline.

THE UNIVERSE, CREATIVE LOVE, AND THE PHYSICISTS

Sir John wondered if Unlimited Love (replete with Creativity and Intelligence) preceded the big bang. This kind of thinking is commonplace in the metaphysical reflections of New Thought and Unity, the Abrahamic traditions, Emanuel Swedenborg (e.g., in his *Divine Love and Wisdom*), and Hinduism. But it is also not all that uncommon among physicists, some of whom Sir John cited with frequency.

For example, Sir John was impressed with Sir James Jeans. In his famous Rede Lecture titled "The Mysterious Universe" delivered at Cambridge University in 1930, the mathematical physicist, an ontological idealist, wrote seriously about the universe "as a world of pure thought" (1943, 175). He wrote. "If the universe is a universe of thought, then its creation must have been an act of thought" (1943, 181). Moreover,

> Time and space, which form the setting for the thought, must have come into being as part of this act. Primitive cosmologies pictured a creator working in space and time, forging sun, moon and stars out of already existent raw material. Modern scientific theory compels us to think of the creator

as working outside time and space, which are part of his creation, just as the artist is outside his canvas. (1943, 182)

For Jeans the Godhead that exists is the Ground of Being before time and space. Jeans wrote further,

> With this caution in mind [that scientific theories change], it seems at least safe to say that the river of knowledge has made a sharp bend in the last few years. Thirty years ago, we thought, or assumed, that we were heading towards an *ultimate reality* of a mechanical kind. It seemed to consist of a fortuitous jumble of atoms, which was destined to perform meaningless dances for a time under the action of blind purposeless forces, and then fall back to form a dead world. Into this wholly mechanical world, through the play of the same blind forces, life had stumbled by accident. (1943, 186, italics added)

But with the 1925 discovery of quantum mechanics, continues Jeans, this mechanical view of Ultimate Reality changed forever:

> Today there is a wide measure of agreement, which on the physical side of science approaches almost to unanimity, that the stream of knowledge is heading toward a non-mechanical reality; *the universe begins to look more like a great thought than like a great machine.* Mind no longer appears to be an accidental intruder into the realm of matter; we are beginning to suspect that we ought rather to hail it as the creator and governor of the realm of matter—not of course our individual minds, but the mind in which the atoms of which our individual minds have grown exist as thoughts. (1943, 186, italics added)

The "old dualism between mind and matter" seems to have evaporated. (1943, 186)

So, as physicist Richard Conn Henry of the Johns Hopkins University wrote in the journal *Nature*, in an article titled, "The Mental Universe," "The Universe is immaterial—mental and spiritual. Live, and enjoy." (2005, 29)

Sir John was sympathetic with the claim that Creative Mind and the energy of Love precede matter. For him an Infinite Intelligence motivated by an essence of Creative Love thinks the universe into existence. A hundred years ago such a perspective would have been laughable, for the universe was viewed for the most part as matter following the laws of Newton. But viewed from a different prism, quantum mechanics, a very different view of reality has developed. We grew up in the second half of the last century studying the atom with the help of marbles and ball bearings. In point of fact, by the 1920s, the simple Newtonian model was abandoned, and with Einstein's theories it became clear that really everything in the universe is made up of energy. Quantum physicists concluded that the atom is not structured like a little solar system. Rather, atoms are more like energy vortices spinning and vibrating, like a toy top, and radiating energy patterns. Atoms are like a tornado vortex, but you can't see them because they don't have all the dirt and debris in them that tornados collect. Now, it seems that there are actually a number of infinitesimally small vortices known as photons and quarks that together form the vague sphere of an atom, and yet the closer you look at the atom with an atomic microscope, the more it disappears from view. In other words, atoms comprise energy rather than matter.

This is a little hard to accept, perhaps, and there is a paradox: matter can be described as immaterial in its essence, yet at the same time, this energy takes on the form of the material reality we touch and feel. Atoms can be seen both ways. Maybe the way to say this is that energy and matter are part of a dynamic whole and not independent. Einstein stated this with the formula energy (E) = matter (m, mass) times the speed of light (c) squared. The idea of a Universal Prime Energy of Unlimited Love from which the big bang burst is not quite as strange an idea as it once might have been.

Of course, this idea was never strange to metaphysical thinkers. Sir John's reflections follow an Americanist tradition that included Charles Fillmore, Ernest Holmes, and Mary Baker Eddy, who all held that "matter may be only an outward manifestation of divine thought, and that the creative spirit called God is the only reality" (1981/1995, 20). This becomes easier to believe, asserts Sir John, in the light of modern physics, which has pointed to the nonexistence of matter since the work of Einstein. (1981/1995, 21)

As the distinguished physicist Kenneth W. Ford notes, a gluon, the "glue" particle within a nucleus between a neutron and a proton, lasts about a billionth of a billionth of a second between its creation and annihilation (2004, 16). Particles like pions and muons and neutrinos pop in and out of existence, and photons break down the distinction between particle and wave. The incessant creation and annihilation of particles gives one the impression that at this quantum level the line between being and nonbeing is fuzzy. Particles come in and out of existence so quickly that they make the blink of an eye look like years. Matter seems to lack solidity and permanence at its quantum level. It is almost ephemeral. Where are these particles coming from? How are they coming into being and out of being? The idea of the universe as grounded in an underlying matrix of Unlimited Love might fit in here somehow, maintaining everything that is.

This intimacy between spirit and matter, and between Creator and creation, points to something deeper than our being made "in God's image" or being "God's children," for in fact we are constantly in God by virtue of our existence (Templeton 1981/1995, 22). God is the universal spirit or energy, the "causative idea which sustains and dwells in all He created and is still creating" (1981/1995, 24). And God does all this "out of love." (1981/1995, 24)

Many physicists have moved away from secularism these days and have been driven by the most widely accepted theories of cosmology and origins to some form of theism that minimally includes an Infinite Intelligence and First Cause, in which the motive of Love seems implicit.

Big Bang Models

Modern science more or less uniformly accepts standard and revised big bang models of the beginning of the universe. Everything in the visible universe is the remnant of a mega-explosion that occurred about

13.7 billion years ago. In this "standard model of cosmology," the universe is expanding through space, and space itself is stretching out. There was in this model an actual beginning to the universe in the strong sense, which includes the beginning of time itself, and of space. Incidentally, long before modern cosmology, Augustine (354–430 CE) suggested that even time had its beginning in a creation event. This idea of the big bang has produced a remarkable openness and interest in the idea of a Designer among premier modern physicists (Owen Gingerich, Fred Hoyle, Arno Penzias, Roger Penrose, John Polkinghorne, and Paul Davies, to list just a very few). Note that this is not the argument for design that those who wish to explain sudden and improbable leaps in biological complexity, and who argue against unguided biological evolution, discuss. Rather, we have in mind here the preponderance of evidence for a beginning to the universe, including space and time and all that is both visible (matter) and invisible (energy). Many physicists, like the theophilosophers of old, cannot explain how something (everything) could come from nothing. There is thus a reasonable likelihood of a creative transcendental First Cause. It is no longer possible to suggest that the universe "just always existed and does not need a cause." It did not always exist; it came into being in a big bang, and it is almost impossible to explain how this came about without some reference to Infinite Intelligence or Divine Mind.

What new understandings will emerge from the subatomic world? Will the idea of an Infinite Intelligence that is the Ground of all these forces and particles and energies that are buzzing and speeding into and out of existence be the only plausible question, and possibly the only plausible answer?

The Mathematical Principles and Constants of the Universe

For many physicists, the fact that we have a mathematically organized universe in which universal laws of mathematics exist and are stable indicates an Infinite Intelligence that created these laws in all their glorious sophistication. This argument seems necessary because where else would such astonishing elaborate principles come from? Moreover, the various universal constants of the universe that hold it all together—

such as the speed of light, the gravitational constant, the strong force constant, the weak force constant, Planck's constant, and so forth—strongly point toward some Infinite Mind that underlies it all. Not only do we have a universe that was created out of nothing, but we have one that is so organized mathematically and with regard to sustaining constants that modern science clearly leaves us with the idea of a First Cause or Transcendent Intelligent Agency as highly plausible and even the most reasonable explanation for the universe as we know it.

The New Physics of Quantum Theory

While Newtonian mechanics and thermodynamics explain much of the action in the visible universe, there is another entire level to the universe at the very smallest levels of being that is mind boggling, to say the least. Essentially, scientists had to abandon the very idea of a material universe as it became clear that matter is really an illusion and that everything in the Universe comprises subtle energies. The physical atom is actually a vortex of energy, spinning and vibrating and wobbly like a top, radiating identifying energy patterns. Quantum physicists discovered that the old Newtonian atomic model with electrons in orbit around protons and the like is simply wrong. Rather, the atom is like a funnel cloud made out of invisible energy rather than tangible matter, and emerging out of nothing. This suggests that Ultimate Reality might be grounded in a Divine Mind and Creative Love from which all energy and, hence, matter emanate.

Sir John studied these theories of physics with care and was in active conversation with leading minds. He stated, "Maybe we will discover that love is indeed the basic force in the spiritual world. Could Dante have been correct when he said, 'It is love which moves the sun and stars'?" (2000a, 163). This question is hard to grasp. It is like thinking of gravity as a benevolent, purposeful force, rather than something that just is. Is such an energy or force as gravity an emanation from an essence of God that is ineffable, unknowable, but love-motivated? Perhaps the divine essence is some sort of pre-energy that precedes all the energies of the universe and the cosmos. At some level, maybe all energy and all the laws of the universe are proceeding forth from the mind of an Infinite Intelligence and an Infinite Heart. Is there a universal prime love from which everything arises? It is tempting to dodge all these big questions about Unlimited Love as Ultimate Reality, for they are so overwhelming, and even asking them would for some people indicate that one has taken leave of one's senses. But surprising numbers of physicists ask these questions in tandem with theologians across all spiritual traditions. *Is God's love a cosmic matrix that keeps everything from falling apart?* Clearly Sir John was fascinated by the very fast-moving discoveries and theories in physics, and he sensed that they would point toward the reality of a consciousness and creative love behind the universe.

Sir John never asserted that modern physics proves God. He was cautionary but wonderfully speculative. He knew that in every generation, some noble arguments are launched using the current physics to prove the existence of a Higher Power. He also understood the minds of the great physicists and would have agreed with Ken Wilber, "According to their general consensus, modern physics neither proves nor disproves, neither supports nor refutes, a mystical-spiritual worldview" (2001, 3). Yet all the great physicists Wilber discusses—Heisenberg, Schrödinger, Einstein, De Broglie, Jeans, Planck, Pauli, and Eddington—encouraged the dialogue between physics and spirituality. Moreover, although none of these figures thought that physics demonstrated God conclusively, none of them thought that physics refutes such a worldview, and most felt that the idea of Infinite Mind was reasonable. Indeed, as Wilber also points out, each and every one of them nevertheless became mystics, and all wrote eloquently of a cosmic religious feeling that informed their lives and their pursuit of science.

THIS ULTIMATE REALITY IS MOST LIKELY LOVE?

Sir John had no tolerance for the destructive images of a wrathful and hateful God that is merely the product of the dark side of human nature and imagination, and against whom the simple-minded atheists rightly rant. No, he held to the perennial philosophy—the physical universe is not the sole reality, human nature includes a nonmaterial soul, and human beings possess a point of contact with the Divine Mind. And especially for Sir John, there are loving purposes in the fine-tuning of the physical laws and constants of nature that are generative of life and evolution.

We can explain the universe in three basic ways.

1. The universe is somehow purpose-guided, and a divine intelligence and love determined the laws and constants with the idea of allowing various life forms to evolve leading to creatures of mind.

2. The universe is all an accident.

3. The universe is a matter of statistics. All sorts of parallel universes exist in a multiverse system, and one universe was fine-tuned just right to give rise to life.

By some estimates this last option would require 10^{500} total universes, more universes than there are atoms in our universe. This makes the idea of an intelligent consciousness as source of our universe seem plausible.

Maybe God more or less dreams or visualizes the universe into existence, with its life-friendly laws and principles that are themselves the result of God's love. These stable laws of thermodynamics, mathematics, gravity, and the like are manifestations of God's preexisting intelligence and creative love as the universe comes into existence through the big bang—a universe in which reality is both visible and invisible, physical and nonphysical, and interchangeable between matter and energy.

Some of Sir John's biggest questions include:

- "Are time and space and energy all part of God, and yet is god much more?"
- "Is god ultimately the only reality—all else being fleeting shadow and imagination from our very limited five senses acting on our tiny brains?"
- "Is god all of you, and are you a little part of him?" (2000a, 20)

Sir John stressed the smallness of our thinking, and he urged us to acknowledge that the images of God we create are human projections:

> Throughout history, have gods created by human minds been too small? God concepts typically follow anthropological typologies. Often, theological constraints are derived by expanding human analogies, as in the very important arguments against the existence of god based on the problem of evil. These considerations are profoundly important for theology and more generally for belief in god. However, they presuppose a framework of analysis based on limited human understanding. Yet, god may work in mysterious ways, ways beyond our limited concepts. Might what appears to us as god's inhumanity be part of a more perfect picture that we simply cannot see? Divine love may not be so small or simplistic as our limited concepts might suppose. (2000a, 20)

Sir John offered this powerful analogy:

> We devise theologies hoping they in some way adequately represent god. And yet they are always inadequate. If god were small enough to fit our human reason, he would not be god at all but a better human. Human comprehension is very limited, just as radio receivers that can receive music, voices and wonderful sounds from hundreds of sources but are hopelessly blind to the sunsets and flowers. Would a god small enough to be fully comprehended by human minds be just a product of human minds? (2000a, 21)

Consider these statements: "God's unlimited love may be the basic reality from which all else is only fleeting perceptions by humans and other transient creatures" (2000b, 3); "Although it is nice to love people, it is because the expression of love—unconditionally and with no limits or restrictions—radiates a great fundamental Invisible reality" (2000b,

14); "Other qualities such as power, wisdom, grace, justice, and so on may be attributes of the Creator's caring for his creatures, but his love is preeminent—and unlimited" (2000b, 18); "Those who are philosophically inclined may find it helpful to understand God's unlimited love as the original and ongoing basic creative force of the universe. This love was present before the beginning, and it continues to hold all things together. Our fleeting human emotions and perceptions are in fact mere glimpses of God's perfect love" (2000b, 19); "So, if there is a phenomenal universal force, for example, in gravity, in the light spectrum, can there not also be a tremendous unknown, or non-researched, potency, or force, in unlimited love?" (2000b, 35–36); "Could unlimited love be described as a creative, sustaining energy? When we embrace our creative energy, can we draw, from the universal Source, a tremendous spiritual energy matrix into many areas of our lives? Does a divine fountainhead of love exist in the universe in which degrees of human participation are possible?" (2003, 87).

These are the kinds of questions that look for a scientific understanding of what the mystics already have experienced. For example, Meister Eckhart (1260–1327), a Christian mystic, referred to both God's outpouring of love in the creation of the universe, and God's actual presence in every soul. As the soul becomes more pure and emptied of material distractions it becomes increasingly aware of the reality of God's presence and love. It is not as though God is standing here and we are over there, because God and we are already one in each individual soul. For Eckhart, God is the Godhead behind all being. God dwells in his image—the soul. In other words, our deepest nature is part of an all-pervading Consciousness in which we are integrated with the Divine Source, and we realize that this eternal aspect of the mind does not come from biology but rather comes from God.

For anyone who has mystically experienced this transforming love energy, it is overwhelming. But what of the objective level of reality? Is Unlimited Love actually an Ultimate Reality, the reality from which all being arises? Sir John asked, "Is agape love a product of the human mind, or can human minds be a product of pure, limitless, timeless love, which some call God?" (2003, 97). "What role might unlimited love play in why you were created and in your personal journey toward purpose?" (2003, 98)

Sir John never shied away from the big metaphysical questions: "Because of the very nature of humanity, we find ourselves in a frustrating dilemma. As finite physical beings we can be filled with self-centered urges. On the other hand, while we may be limited, we may live as a part of spiritual infinity—*a presence*—that impels us to reach for the highest spiritual ideals. What is this spiritual infinity?" (2000b, 34). He could ask, boldly, "So, if there is a phenomenal universal force, for example, in gravity, in the light spectrum, or in electromagnetism, can there not also be a tremendous unknown, or non-researched potency, or force, in unlimited love?" (2000b, 35–36)

He made an analogy between the visible sun and the invisible spiritual energy of Unlimited Love:

> The sun of our solar system has been described as a self-sustaining unit whose energy source is derived from internal thermonuclear reactions. Scientists tell us the energy released in these reactions is so vast that the sun could shine for millions of years with little change in its size or brightness. Does an analogy between the strength, power, light, warmth, and wisdom of unlimited love correlate with the radiance of our solar system's sun? Especially with the knowledge that invisible electromagnetic energy fields surround all living things. Both the sun and unlimited love bring life-giving warmth and light to a planet and its people. (2003, 87)

Consider that we take the energy of the sun for granted day by day because it is the background reality of our lives and of life itself. Is it possible that there is an energy of Unlimited Love that is completely essential and basic to all being and interaction, but we take it so much for granted that most of us are entirely unaware of it, however much we would all be abruptly aware of its sudden absence? Sir John asked,

> Could unlimited love be described as a creative, sustaining energy? When we embrace our creative energy, can we draw, from the universal Source, a tremendous spiritual energy matrix into many areas of our lives? Does a divine fountain-head of love exist in the universe in which degrees of human participation are possible? (2003, 87)

Perhaps if we could open our perception to it, we would see that we are already completely sustained by Unlimited Love, even though our senses may tell us that we are not. After all, we think we are sitting still in a chair, when in fact we are flying around at thousands of miles an hour because of the earth's revolution around the sun.

Sir John liked to use the word *manifest,* and he saw God as manifest in the created world at the most intimate sustaining levels:

> God is infinite. Everything that exists in the universe and beyond the universe is God. That means that the visible universe is only a small particle of God and is itself a manifestation of God. (1987, 130)

By the word *manifest,* Templeton means that which a human being can know. Thus one little particle of God has become known to us through gravity, light waves, pulsars, and other things that enable us to perceive a few features of the universe.

"Matter and energy," he writes, may be "only contingent manifestations of God," and God is their constant sustainer (1981/1995, 21). In short, "Nothing exists in separation from God" (1981/1995, 20). To state that Ultimate Reality may be Unlimited Love may sound preposterous to some. But the fact is that, "Up until this century people would have said that this table is reality. But now natural scientists know that it is in fact 99 percent nothingness. What appears as reality in your eyes is just a configuration, a constantly changing vibration. What we conceive of as reality is really appearance." Moreover, "The only reality is the Creator. He and his works are the only permanent things. I would put it this way: The things that are unseen are reality. The illusions, the temporary things, are what we see" (1987, 147).

Sir John could only imagine a creative loving mind and energy underlying all the laws, order, and thermodynamics of the universe that "set up" (the "anthropic principle") evolutionary "complexification" and a creature capable of knowing God's love. He did not claim that this was more than a plausible hypothesis at this time:

> Certainly, there seems to be no conclusive argument for design and purpose, but there are strong evidences of ultimate reality more fundamental than the cosmos. So, if there are phenomenal universal forces, for example, in gravity, in the light spectrum, or in electromagnetism, can there not also be a tremendous unknown or non-researched potency or force of unlimited love? With earthly information now doubling every three years, can our comprehension of some of these intangibles of spirit also be multiplied more than one hundredfold? Could unlimited love also be an aspect of dimensions beyond what we presently know as time and space? Could unlimited love be a universal concept beyond matter and energy as they are currently understood? To what realms beyond the physical might unlimited love reach? Just how vast is the reach of unlimited love? (2003, 95)

Is God the creative love energy and intelligence underlying all things visible and invisible? Is God the *substance* (from *sub*, under, and *stance*, stand) or Ultimate Reality standing underneath all of reality, or the literal Ground of Being? Just like we can plant our feet on the ground and reflect on the thousands of miles of substance below them, so God may be the deepest source of all Being, ever re-creating and sustaining, ever renewing in Unlimited Love all that is.

Chapter Nine

Three Points of Evidence for Unlimited Love as Ultimate Reality

IN 1893 the Harvard philosopher Charles S. Pierce proclaimed "the great evolutionary agency of the universe to be Love" (1955, p. 361). He distinguished this agency from *eros* or passion, and instead described it as a "cherishing love." Referring to the Gospel of John, Pierce wrote, "Nevertheless, the ontological gospeller, in whose days those views were familiar topics, made the One Supreme Being, by whom all things were made out of nothing, to be cherishing love" (p. 361). "Ontology" refers technically to "being," and Pierce's point is that God in the form of Divine Love underlies everything that has being including the very universe itself.

Like Pierce, Sir John Templeton was an "ontological gospeller." He assumed that God's "cherishing" love is the great active agent underlying our universe. This idea has been diminished by the acids of modernity and secularism, and some readers will deem it archaic. However, it is an idea that has had a tremendous renaissance over the last twenty years as so many thoughtful spiritual thinkers have articulated it anew, particularly at the interface between science and spirituality. Sir John drew on the perennial "ontology" of the Gospel of John repeatedly, and he was ahead of his time as he looked to a dialogue with science that might in future decades resuscitate ancient spiritual wisdom and make it more relevant and digestible for those who have given up on deep questions.

In the previous chapter I presented some of Sir John's basic thoughts as "ontological gospeller." Now I proceed to three areas identified in his writings and conversations with me where he thought evidence for Unlimited Love as Ultimate Reality might be mined fruitfully. This chapter covers these in some depth.

EVIDENCE I: THAT PEOPLE WIDELY SELF-REPORT SPIRITUAL EXPERIENCES OF UNLIMITED LOVE AND AN ASSOCIATED ENHANCEMENT OF BENEVOLENCE

The first indication that Unlimited Love is a spiritual reality must be that it is widely self-reported as experienced, and that its active effects are observed behaviorally in amplified benevolent actions. True, this does not prove the reality of Unlimited Love, but it is one leg of a three-legged stool that provides support.

In the future some sort of device might be invented to detect a mysterious Unlimited Love energy, as people do report the sensation of such an energy field. Sir John actually once suggested this to me. Indeed, in his August 3, 2001, letter to me he wrote, "Can methods or instruments be invented to help humans perceive larger love, somewhat as invention of new forms of telescopes helps human perceptions of the cosmos?" Methods are easier, "instruments" more challenging. But maybe someday someone will invent some sort of device to detect God's love energy flowing through high givers. Seriously, if God really is love, then something might one day be invented to pick up Unlimited Love energies.

Not having such a device to measure Unlimited Love, the best we can do is investigate human experience as subjectively reported, and consequent patterns of behavior. One self-report of such an experience comes from the poet W. H. Auden, and here I repeat for emphasis just a few lines from his description of a quiet experience of God's Unlimited Love as he described it:

> One fine summer night in June 1933 I was sitting on a lawn after dinner with three colleagues, two women and one man. We liked each other well enough but we were certainly not

intimate friends, nor had anyone of us a sexual interest in another. Incidentally, we had not drunk any alcohol. We were talking casually about everyday matters when, quite suddenly and unexpectedly, something happened. I felt myself invaded by a power which, though I consented to it, was irresistible and certainly not mine. For the first time in my life I knew exactly—because, thanks to the power, I was doing it—what it means to love one's neighbor as oneself. (1965, 30–31)

This experience of being "invaded" speaks for itself, and the experience is not terribly uncommon, as we shall see.

Another example is that of Bill Wilson, cofounder of Alcoholics Anonymous, not previously a believer, who claims he saw and felt a white light that he perceived as the presence of God's love. This occurred in his hospital room at a New York detox center on his fourth day of treatment: "It seemed to me, in the mind's eye, that I was on a mountain and a wind not of air but of spirit was blowing. And then it burst upon me that I was a free man" (as cited in Brooks 2010, A31). Bill W. never drank again after that spiritual experience of December 14, 1934. It was an experience of transforming love for a man who felt that he deserved none. Bill W. went on to develop Alcoholics Anonymous into a worldwide organization that would save countless millions of lives.

Just imagine! What if a great majority of individuals responding to a randomized social scientific population survey report that they have indeed had an overwhelming feeling of being invaded by an uplifting spiritual energy of unconditional love that seems to catch them by surprise, like a sudden spring breeze from nowhere that invades the soul and melts the ice? What if fewer than one out of every five individuals have *not* had such an experience? This is indeed the case in the United States, at least. Could there really be an emotional energy field of Unlimited Love? Can we leap into this like electrons jumping into higher orbits if we prime the experience by behaving in loving ways? Metaphysics aside, the social fact is that the experience is widely reported.

In 1900 the psychologist and philosopher William James was busy investigating this aspect of human experience by analyzing existing spiritual writings as they have accumulated over the centuries in various cultures and parts of the globe. His book *The Varieties of Religious Experience* remains a thick descriptive classic that left James concluding that there may likely be some "More" in the universe that underlies these experiences. James wrote that "ripe fruits of religion" are universally understood in terms of saintliness, which includes these features: a sense of the existence of an Ideal Power, a self-surrender to its control, a sense of elation and freedom, and "a shifting of the emotional centre toward loving and harmonious affections, towards 'yes, yes,' and away from 'no,' where the claims of the non-ego are concerned" (1902/1982, 272–73). Religious experience, in relation to charity, is described thus: "The shifting of the emotional centre brings, secondly, increased charity, tenderness, for fellow-creatures. The ordinary motives to antipathy, which usually set such close bounds to tenderness among human beings, are inhibited. The saint loves his enemies, and treats loathsome beggars as his brothers" (1902/1982, 74). Moreover, "Brotherly love would follow logically from the assurance of God's friendly presence, the notion of our brotherhood as men being an immediate inference from that of God's fatherhood of us all." (1902/1982, 78).

However, James never thought to perform a national survey. Survey studies provide a broader picture of the frequency and extensiveness of such experiences across a population. In this discussion I briefly describe a scientific survey of randomly selected Americans conducted in 2010 with my two sociological colleagues, Matthew T. Lee and Margaret M. Poloma. This presentation is brief because all the details are available in our book titled *The Heart of Religion* (Oxford University Press, 2013). The survey respondents were adult (eighteen years of age or older) and selected regardless of religious background, economic strata, educational level, ethnicity, or any other factor. Thus, the survey was totally random, designed to provide a scientifically valid portrait of the experience of God's love and its ramifications.

Select figures from our 2010 Godly Love National Survey (GLNS), commissioned by the Flame of Love (FOL) Project with support from

the John Templeton Foundation, give the reader a sense of the importance of these phenomena within the context of American culture. The
United States is more religious (regardless of whether the dimension
is private devotion, public ritual, or religious experiences) than other
Western countries, including Canada, and we have not extended out
survey to other countries or regions of the world.

Our national telephone survey was open to all American adults
whether they were religious or not. We collected a random sample
involving 1,208 American adults (men and women; across the spectrum
of age, race and ethnicity, geographic location, income, education, etc.).
Respondents were interviewed by telephone in English or Spanish in the
fall of 2009. The results can be generalized to the vast majority of Americans, with a margin of error of plus or minus 2.9 percentage points.
The survey was conducted by the Bliss Institute of Applied Politics at
the University of Akron under the direction of Lee, Poloma, myself,
and John C. Green, one of the nation's leading survey methodologists.

Our survey reveals these results in response to the question, "Do you
feel God's love for you directly?"

Never/not asked	17.4 percent (N=210)
Once in a while	13.0 percent (N=156)
Some days	10.5 percent (N=126)
Most days	14.1 percent (N=170)
Every day	35.6 percent (N=427)
More than once per day	9.3 percent (N=112)

This adds to 99.9 percent due to rounding (N=1,202 completed
responses, with 7 nonresponses). The survey had 1,209 respondents,
but only 1,202 answered this particular question (4 refused and 3 indicated that they didn't know or couldn't remember). In terms of the sub-
stantive importance of the experience of divine love for benevolence,
findings from this survey showed that the 9% (N=112) who feel God's
love more than once per day are the highest givers of time, energy, and
money in service of the neighbor.

Eighty-one percent of Americans acknowledge that they "experience

God's love as the greatest power in the universe," and 83 percent said they "feel God's love increasing their compassion for others." Those who feel God's love more than once per day are more than twice as likely as the rest of Americans to give their time to help others in need, and more than twice as likely to give more than $5,000 per year to help others in need. They are also more likely to help at the widest level of extensivity (at the world level). In multivariate analysis, Divine Love was the only significant predictor of all six of our measures of benevolent service, independent of commonly used controls. We created scales based on multiple questions in order to measure benevolent behaviors (not just attitudes or beliefs) at different levels of extensivity (friends, family, community, and world). Never before has this level of detail about the experience of Divine Love been collected. After controlling for other factors, those who claim to experience God's love most often are the ones who are most generous with their time and money and most extensive in their benevolent service to others (beyond the near and dear).

Let us reiterate in slightly different terms for amplification: Almost half (45 percent) of all Americans feel God's love at least once a day and eight out of ten have this experience at least "once in a while." Nine percent claim that they experience God's love more than once a day. Only 17 percent report no experience of God's love. Eighty-three percent indicate that they "feel God's love increasing their compassion for others." Thus, millions of Americans frequently experience Divine Love, and for them this not only enhances existential well-being, but underlies a sense of personal meaning and purpose and enlivens compassion for others.

Do spirituality, religiosity, regularly sensing God's love, and existential well-being work together to contribute to increased concern for the well-being of others? We used a statistical procedure known as multiple regression analysis in an attempt to tease out which variables best accounted for agreement with the statement that "God's love has increased my compassion for others." In the order of importance (beginning with the largest beta score) the variables are *importance of religion* (.34), *experiencing God's deep personal love* (.27), *importance of spirituality* (.19) and *having a strong purpose in life* (.07). (The adjusted R square was .50.)

Beneath our culture's obsession with wealth and power, status and celebrity, millions of Americans are quietly engaged in a deeply religious struggle to wake up from petty selfishness and to embrace a life of benevolence and compassion.

So is Unlimited Love Ultimate Reality? Eight out of ten Americans claim to have experienced God's love, and they consider it to be "the greatest power in the universe." *While the idea that God's Unlimited Love is Ultimate Reality has not been proven from the perspective of the physical sciences, in the minds of most Americans based on personal experience, this mystery is a social fact.*

A Model to Distinguish Divine Unlimited Love from Mere Human Love

We cannot leave this first area of evidence without emphasizing that Unlimited Love implies an elevated level of consciousness of the Divine and is not simply a dimension of the human substrate. When Sir John considered Unlimited Love, he was not thinking of mere human love. The discontinuities between pure Unlimited Love and human love are obvious to anyone who reads the newspapers. He cited Pitirim A. Sorokin (1889–1968), a towering figure in twentieth-century sociology, who addressed the need to affirm a "supraempirical" source of love in his classic work, *The Ways and Power of Love* (1954/2002). Sorokin described a first dimension of love—its *intensity*. Low-intensity love makes possible minor actions, such as giving a few pennies to the destitute or relinquishing a bus seat for another's comfort; at high intensity, much that is of value to the agent is freely given. Human love can be intense or weak at any given moment, for we are subject to the ups and downs of the heart. Divine love is *perfectly intense*, meaning that its energies are infinite and constant. Human love falls short.

Sorokin's second dimension of love is *extensivity*: "The extensivity of love ranges from the zero point of love of oneself only, up to the love of all mankind, all living creatures, and the whole universe. Between the minimal and maximal degrees lies a vast scale of extensivities: love of one's own family, or a few friends, or love of the groups one belongs to—one's own clan, tribe, nationality, nation, religious, occupational, political, and other groups and associations" (1954/2002, 6). Our human love tends toward insularity and has a strongly myopic tendency unless it is enlarged by the power of Divine Love.

Sorokin next added the dimension of *duration*, which "may range from the shortest possible moment to years or throughout the whole life of an individual or of a group" (1954/2002, 6). For example, the soldier who saves a comrade in a moment of heroism may then revert to selfishness, in contrast to the mother who cares for a sick child over many years. Romantic love, he indicates, is generally of short duration as well. Human love is notoriously fickle. It gives up on people when they have been wounded by life, or when they fail to measure up to some standard of achievement. Divine Love is perfectly reliable and enduring; human love is not.

The fourth dimension of love is *purity*. Here Sorokin wrote that our love is characterized as affection for another that is free of egoistic motivation. By contrast, pleasure, advantage, or profit underlie inferior forms of love, and will be of short duration. Pure love—that is, love that is truly disinterested and asks for no return—represents the highest form of emotion. Human love is at best a mix of other-regarding and self-regarding motives, whereas Divine Love is pure.

Finally, Sorokin included the *adequacy* of love. Inadequate love is subjectively genuine but has adverse objective consequences. It is possible to pamper and spoil a child with love, or to love without practical wisdom. Adequate love achieves ennobling purposes and is, therefore, anything but blind or unwise. Certainly love is concerned with the building of character and virtue, and will shun overindulgence. Successful love is effective. Human love is sometimes very destructive in its application. We destroy our children by loving them in ways that breed bad habits and irresponsibility. Divine Love is perfectly creative because it is perfectly wise.

Sorokin argued that the greatest lives of love and altruism approximate or achieve "the highest possible place, denoted by 100 in all five dimensions," while persons "neither loving nor hating would occupy a position near zero" (1954/2002, 6). Gandhi's love, for example, was intensive, extensive, enduring, pure, and adequate (effective). Of special interest to Sorokin was the love of figures such as Jesus, Al Hallaj, and Damien the Leper. Despite being persecuted and hated, and therefore without any apparent social source of love energy, they were nevertheless able to maintain a love at high levels in all five dimensions. Such love seems to transcend ordinary human limits, which suggested to Sorokin that some human beings do, through spiritual and religious practices, participate in a love energy that defines God.

Sorokin was convinced that such perfect love can best be explained by *hypothesizing an inflow of love from a higher source that far exceeds that of human beings*. Those who were despised and had no psychosocial inflow of love to sustain them must receive love from above:

> The most probable hypothesis for them (and in a much slighter degree for a much larger group of smaller altruists and good neighbors) is that an inflow of love comes from an intangible, little-studied, possibly supraempirical source called "God," "the Godhead," "the Soul of the Universe," the "Heavenly Father," "Truth," and so on. Our growing knowledge of intra-atomic and cosmic ray energies has shown that the physico-chemical systems of energies are able to maintain themselves and replenish their systems for an indefinitely long time. If this is true of these "coarsest" energies, then the highest energy of love is likely to have this "self-replenishing" property to a still higher degree. We know next to nothing about the properties of love energy. (1954/2002, 26)

Sorokin believed that people of great love who sustained that love when surrounded by adversity were graced. Those who are high in all five aspects of love reflect, he conjectured, a divine love energy.

Sir John was excited by Sorokin because he too did not want to collapse God's Unlimited Love into mere human love. Sir John would agree with what Sorokin describes as a "still higher level in the mental structure of man, a still higher form of energies and activities, realized in varying degrees by different persons—namely, the supraconscious level of energies and activities" (1954/2002, 96). The perfectly integrated creative genius achieves the highest level of creativity without strenuous effort. Sorokin gathers empirical support for this statement from the testimony of "innumerable eminent apostles of love" who, across cultures and generations, describe themselves as instruments of the supraconscious: "God, Heaven, Heavenly Father, Tao, the Great Reason, the Oversoul, Brahma, Jen, Chit, the Supre-Essence, the Divine Nothing, the Divine Madness, the Logos, the Sophia, the Supreme Wisdom, the Inner Light." (1954/2002, 127)

EVIDENCE 2: THAT PEOPLE WHO LOVE GOD AND THEIR NEIGHBORS AS THEMSELVES GENERALLY LIVE HEALTHIER AND MORE JOYFUL LIVES

Sir John Templeton often cited the double love commandment of Matthew 22 as follows:

> Thou shalt love the Lord thy God with all thy heart, and with all thy soul, and with all thy mind. This is the first and great commandment. And the second is like unto it, Thou shalt love thy neighbor as thyself. On these two commandments hang all the law and the prophets. (2000a, 160)

Following the major spiritual traditions, Sir John prescribed a triadic relational nexus between a Higher Power, self, and others (neighbor), consistent with what Aldous Huxley called the "perennial philosophy" (1945), articulated among others by Jesus of Nazareth who, drawing on Jewish tradition, taught that human fulfillment is optimal when we creatively adhere to the double love commandment. Here God, neighbor, and self are interwoven in a triadic community of being that defines what I have termed the *ontological generality* (Post 2012). By this I mean that at the very deepest level of our being, we are each oriented toward God, neighbor, and self. Abiding in this triadic structure allows true fulfillment. Is this true? If it is so, then it provides another leg on the tripod of evidence for Unlimited Love as Ultimate Reality.

Fulfillment is a unique word. It comes from the Old English *full fyllan*, or literally to "fill up fully" or to make "fully filled." Its opposite is emptiness, and a related restlessness. The term *fulfillment* allows for an extraordinary dialogue between theological and psychological discourse ranging from Augustine to Victor Frankl, from Pascal to Huxley. In Ephesians 3:17–19, Paul hoped that his fellow believers "may be filled to the measure of all the fullness of God."

The fulfillment that comes from living in accordance with the double love commandment is captured in the Hebrew word *shalom*, which refers to a profound inner peace, well-being, and wholeness that comes from loving God with all our heart and soul, as well as our neighbor as self. When our loves are properly ordered within this triadic framework, we find this fulfillment inevitably because this is how our very being is oriented at the spiritual level. Following Kierkegaard, in his book *Works of Love* (1847), there is nothing wrong with love of self, but the problem is that we love ourselves wrongly; in other words, outside of the triadic structure we do not love ourselves rightly.

Our institute experiment with regard to ontological generality focuses on recovery from alcoholism and Alcoholics Anonymous (A.A.) because here the recovery process is said to occur best at the intersection of two axes: (a) the human axis of love of neighbor and of self, and (b) the divine axis of love for God and of God's love for self and other. Are the benefits to the alcoholic of serving others (the twelfth step) greater in intensity and duration when (a) and (b) converge in the life of the agent, as definitive of the ontological generality? Does the agent's self-reported experience of a Higher Power increase as he or she becomes increasingly diligent in contributing to the lives of other alcoholics? Does the agent's love of self increase when (a) and (b) converge, manifesting in better self-care and stewardship, in part because the alcoholic deems him- or herself to be more valuable in this triadic context than outside of it?

A.A. is the ideal context for the study of the ontological generality in relation to health and fulfillment. The book *Alcoholics Anonymous*, subtitled *The Story of How Many Thousands of Men and Women Have Recovered from Alcoholism*, is called the "Big Book" in A.A. circles. First printed in 1939 (now in its 2001 fourth edition), the opening segment of this spiritual-moral treatment manual begins with the words, "We of Alcoholics Anonymous." The essence of the program is captured in the passage, "we work out our solution on the spiritual as well as an altruistic plane" (1939/2001, xxvi). And from this triadic framework of two planes a new fulfillment becomes possible, including many core spiritual virtues and ultimately sobriety itself.

On the horizontal axis (or "plane") of self and neighbor, the idea that helping others might have benefits to the agent is also a part of American consciousness. It is fairly well established that the horizontal axis has clear benefits. The year 2010 was exciting for research on health, happiness, and helping others. This author was able to consult for the United Healthcare/Volunteer Match Do Good Live Well Study (see www.dogoodlivewell.org/UnitedHealthcase-VolunteerMatch-Do-GoodLiveWell-Survey.pdf), an online survey of a national sample of 4,582 American adults 18 years of age and older. The survey was conducted by TNS (Taylor Nelson Sorfres), the world's largest custom survey agency, from February 25 to March 8, 2010. These remarkable facts stand out:

Forty-one percent of us volunteered an average of one hundred hours per year in 2009 (males, 39 percent; females, 42 percent; Caucasian, 42 percent; African American, 39 percent; Hispanic, 38 percent), and 69 percent of us donate money.

Sixty-eight percent of volunteers agree that volunteering "has made me feel physically healthier," 92 percent that it "enriches my sense of purpose in life," 89 percent that it "has improved my sense of well-being," 73 percent that it "lowers my stress levels," 96 percent that it "makes people happier," 77 percent that it "improves emotional health,"

78 percent that it helps with recovery "from loss and disappointment."

Volunteers have less trouble sleeping, less anxiety, less helplessness and hopelessness, better friendships and social networks, and a sense of control over chronic conditions.

Twenty-five percent volunteer through workplace, and 76 percent of them feel better about their employer as a result.

Coupled with our national survey of the experience of God's love, the findings from United Healthcare suggest that, for large numbers of Americans, the "spiritual as well as an altruistic plane" working in combination and simultaneously for the good of the agent as a side effect of authentic love appears to be an understandable dynamic.

A.A. is an experiment in the ontological generality. Addiction is the breakdown of the triadic community between God, neighbor, and self. Recovery involves developing or restoring this community. The twelve steps of A.A. assert that little good can happen in the life of an alcoholic until a community is established between self, other, and God.

Nowhere is the word "I" to be found in A.A.'s key passage because self-preoccupation is considered the root of the problem. Grandiosity is replaced by anonymity and humility. Any solution lies in the "we" of fellowship centered on a Higher Power, and the recognition that "I" cannot rescue myself (A.A. 1939/2001, 201). As the "Big Book" emphasizes, "Selfishness—self-centeredness! That, we think, is the root of our troubles" (1939/2001, 62). We must be rid of this by becoming "less and less interested in ourselves, our little plans and designs" (1939/2001, 63), and more interested in what we can "contribute to life" (1939/2001, 63). Moreover, "Our very lives, as ex-problem drinkers, depend upon our constant thought of others and how we may help their needs" (1939/2001, 20). Still, our helping others in need must be based in "a sincere desire to be helpful" (1939/2001, 18). All of this prosociality, however, is clearly positioned under the sacred canopy of a Higher Power.

This spirituality achieves several important things. First, such a Higher Power functions to create an absolute quality to abstinence, which becomes more than a mere human contrivance or a matter of "relative" value. Abstinence is therefore nonnegotiable. Second, reliance on a Higher Power takes the place of alcohol in filling the emptiness or incompleteness within. This theme of spiritual emptiness and the misplaced efforts to find fulfillment through things other than God's love can be found in the perennial wisdom of most spiritual traditions. Third, this spirituality frees the self to concentrate on contrition and service. Fourth, the spirituality of A.A. is completely democratic and nonhierarchical, resembling the open polity of Quakerism and other spiritual movements emphasizing the equal status of all members.

Progress is made by the daily pruning of egocentrism: "Selfishness—self-centeredness! That, we think, is the root of our troubles" (1939/2001, 62). Further, "Above everything, we alcoholics must be rid of this selfishness. We must, or it kills us!" (1939/2001, 62). The "Big Book" refers to selfish resentment, dishonesty, self seeking, and unkindness, among other manifestations. Prayer and meditation are prescribed as spiritual practices necessary to remain "in contact" with a Higher Power and the will of God for our lives. The twelfth step, "Having had a spiritual awakening as the result of these steps, we tried to carry this message to alcoholics, and to practice these principles in all our affairs" (1939/2001, 60) is vital. The "Big Book" is abundantly clear: "Our very lives, as ex-problem drinkers, depend upon our constant thought of others and how we may help meet their needs." (1939/2001, 20)

A.A. literature speaks of these as four interwoven principles: surrender, reliance on a Higher Power, redemption, and service. Each principle constitutes a transition away from egocentrism to a connectedness with a Higher Power and with others that constitutes an expansion of being in the direction of the ontological generality. Overall, the twelve steps constitute a technology of spiritual transformation that implements what the great Jewish thinker Martin Buber described as a shift from one way of being in the world ("I-It") to a better way ("I-Thou"). According to the first, "I" relate to others only insofar as they contribute to my little agendas; according to the second, I relate to others as valuable in themselves. Clearly in the modality of "I-Thou" the self flourishes in a way that it cannot under the dictates of selfishness.

The Ontological Generality as Prototype

Is the ontological generality of A.A. implementation an archaic artifact to be buried in the name of progress? Or does it reveal an aspect of human "whole" fulfillment that is, rather, a prototype for the future of self-care in an era when many of the spiritual-religious traditions that explicitly teach self-stewardship have atrophied (Sorokin 1954/2002)? Can the ontological generality that is so successful in A.A. come to the rescue of a great many lives that are lived in what Henry David Thoreau termed "quiet desperation"? I do not just have in mind other illness groups, but people generally speaking. Might the ontological generality be the ultimate key to preventive medicine? Imagine how much destructive and self-destructive behavior could be avoided. The Seventh Day Adventists, for example, are the longest-lived and healthiest subpopulation in the United States, as are spiritually active Jews. Both traditions include the ontological generality and associated teachings on diet, hygiene, and care of the self. Might A.A. be a signpost toward health that our modernized secular societies could benefit from, broadly considered?

There are many heuristic keys into why A.A. is effective for those individuals who decide, on the basis of an elective affinity, to stick with it rather than contribute to the considerable attrition rate during the first year. No shoe fits all, and clearly there is a period of just trying A.A. on for size. Some people react against A.A. almost immediately because they are highly secular, not interested in this or any form of "group therapy," or worried about the rigidity of these mysterious twelve steps. There is an intensity about the A.A. fellowship that does not resonate with the strict individualist who would prefer one-on-one therapy. In many ways, joining A.A. on a sustained and deeply immersed basis is like joining a new religious movement, but without the hierarchy or cult of personality. It might be equated with a sectarian movement, then, with a high esprit de corps.

For those who are freely drawn to the hospitality, enthusiasm, and twelfth-step helpfulness of its core membership, and who wish to commit to A.A. at a serious level of longer-term engagement, there is always the possibility of a relapse, but in general these individuals sustain their sobriety relatively well over the decades. In case a member does relapse and need hospitalization, his or her A.A. peers will make visits in the hospital rehab unit, showering that member with conversation and encouragement.

What about A.A. is doing the work of recovery? I assert that it is the ontological generality, which always includes both a horizontal (self and other) component and a vertical (self and God) component (Levin and Post 2010).

Most long-term members of A.A. do testify to some reconciliation with a Higher Power (aka "God," "Ultimate Reality," "Ground of Being," "The Absolute") that fulfills them and displaces the need for alcohol. In A.A., healing occurs through reliance on a Higher Power, however understood. Spirituality and prayer are clearly operative in A.A. along each of the steps. Divine Love is often experienced in the form of a sense of being accepted unconditionally by a Higher Power who forgives, and this in turn encourages members of A.A. to pass on such acceptance to their fellow alcoholics.

One aspect of the ontological generality is, as mentioned, care of the self. We live in a culture and a time when the care of the self is foundering at many levels outside of the love of God and neighbor. It may be difficult to care for self without reasons beyond the self to do so. Care of the self within the ontological generality is a stewardship that is grounded in a deep appreciation for the love one receives from God and others, which so greatly enhances one's sense of significance and dignity. Why abide by healthy lifestyle behaviors and refrain from self-harm if one's life is focused merely on self? The self is not meaningful enough to care for itself. Salutogenic meaning, in the deep sense that one's life is more than an exercise in fleeting emptiness, is found in the mutual love of the ontological generality.

Spiritualities and religions can enhance health and prevent disease through the care of the self (e.g., self-control in diet and sexual activity, the eradication of smoking and substance abuse, physical exercise and other positive health practices, nonviolence), but this is care grounded in the ontological generality. We have referred to the Seventh Day Adventists, who are particularly long-lived, and to certain Jewish populations. To select a purely symbolic number, let us say that 95 percent of the health benefits of spirituality and religion have to do with care of the self. When looked at globally, this is probably the most significant worldwide contribution that spirituality and religion make to the human condition.

We need a global program in spiritual flourishing and the health of body, mind, culture, and society. Let us reconceptualize and re-create the field anew at a global preventive level. Can we really begin to understand how certain spiritualities and religions promote health in certain regions of the globe? Is there a future in which the value of care of the self and care of the other under a sacred canopy can be much better appreciated, cultivated, and acknowledged? Prevention and responsibility are the future.

Has there been a decline in self-care in the United States, or in other countries across the globe? What are the deeper spiritual, cultural, individual, and community underpinnings of good care of the self? What is the history of self-care? Have the traditions of self-care broken down? Do any features of modern society work against self-care? Is the ontological generality our hope for a paradigm shift that can bring down health-care costs through prevention and self-care?

Maria I. Pagano has led the study of helping behaviors of alcoholics with a range of sixteen to twenty-five years of continuous abstinence from alcohol. While helping others in general was rated as significant in contribution to sobriety, considerably higher benefits came from increased helping of other alcoholics in the context of A.A. (Pagano et al. 2009). Earlier, Pagano and colleagues (2004) examined the relationship between helping other alcoholics to recover (the twelfth step) and relapse in the year following treatment. The data were derived from a prospective study called Project MATCH, which examined different treatment options for alcoholics and evaluated their efficacy in preventing relapse. Two measures of helping other alcoholics in Alcoholics Anonymous (being a sponsor and having completed the twelfth step) were isolated from the data. Proportional hazards regressions were used to separate these variables from the number of A.A. meetings attended during the period. The authors found that "those who were helping were significantly less likely to relapse in the year following treatment." Among those who helped other alcoholics (8 percent of the study population), 40 percent avoided taking a drink in the year following treatment; only 22 percent of those not helping had the same outcome. Imagine, helping others doubles the likelihood of recovery from alcoholism in a one-year period!

Obviously, living under the sacred canopy of the ontological generality moves the self away from narcissism, solipsism, and sin. "I" becomes less important than "Thou." In a study that goes back to 1983, Larry Scherwitz and his researchers at the University of California analyzed the speech patterns of 160 type-A personality subjects (i.e., always in a hurry, easily moved to hostility and anger, high levels of competitiveness and ambition). His data showed that the incidence of heart attacks and other stress-related illnesses was highly correlated with the level of self-references (i.e., "I," "me," "my," "mine," or "myself") in the subject's speech during a structured interview. High numbers of self-references significantly correlated with heart disease, after controlling for age, blood pressure, and cholesterol (Scherwitz et al. 1983). The researchers suggested that patients with more severe disease were more self-focused and less other-focused. The researchers recommend that a healthier heart can result when a person is more giving, listens attentively when others talk, and does things that are unselfish. Something about being self-obsessed or self-preoccupied seems to add to stress and stress-induced physical illness.

Ralph Waldo Emerson, in his famous essay on the topic of compensation, wrote, "It is one of the most beautiful compensations of this life that no man can sincerely try to help another without helping himself." The sixteenth-century Hindu poet Tulsidas, as translated by Mohandas K. Gandhi, wrote, "This and this alone is true religion—to serve others. This is sin above all other sin—to harm others. In service to others is happiness. In selfishness is misery and pain." The ninth-century sage

Shantideva wrote, "All the joy the world contains has come through wishing the happiness of others." Proverbs 11:15 reads, "Those who refresh others will be refreshed." Martin Buber described the moral transformation of shifting from "I-It" to "I-Thou," from a life centered on self as the center of the universe around whom, like the sun, all others revolve. This "I" relates to others only as means to its own ends. But the spiritual and moral self of "I-Thou" discovers "the other as other," and relates to them in compassion and respect. There is still an "I," of course, but a deeper and better "I"; science now shows a happier and healthier "I" as well. Every major religion recommends the discovery of a deeper and more profound human nature, designated in various ways as the "true self." In Acts 20, we find the words, "'Tis better to give than to receive," words that echo down into the Prayer of St. Francis. Now science says it's so.

The ontological generality has been described as perennial, which is to state that it does not and will not go away. Sir John Templeton once wrote, "Our souls long for relationship with God—by whatever name we call the Creator of all there is" (2003, 6). This persistent longing of human nature is now the subject of various explanatory models. Pascal Boyer, for example, advances an argument for the permanence of a religious inclination that is grounded in sociology and evolutionary psychology. His *Religion Explained* challenges the positivist's assumption that belief in a Creative Presence could be set aside in the human future (2001). Andrew Newberg describes the ways in which the human brain appears "hard-wired" for spiritual and religious experiences in his work titled *Why God Won't Go Away: Brain Science and the Biology of Belief* (2001). While views of human nature vary, especially with regard to what is essential or inessential to it, these scientific works demonstrate with varying degrees of success that spirituality may well be ingrained in human nature.

We do see periods in which certain intellectual circles set these sorts of speculations aside. But as Huston Smith argues, questions about ultimate reality and ultimate meaning are grounded in humanness and resist abolition:

Wherever people live, whenever they live, they find themselves faced with three inescapable problems: how to win
food and shelter from their natural environment (the problem
nature poses), how to get along with one another (the social
problem), and how to relate themselves to the total scheme
of things (the religious problem). If this third issue seems less
important than the other two, we should remind ourselves
that religious artifacts are the oldest that archeologists have
discovered. (2001, 211)

Smith describes a modern "tunnel" that attempts to suppress big-pic-
ture thinking: the floor is scientism, the left wall is higher education, the
roof is the media, and the right wall is the law. But the human rational inclination to raise metaphysical questions cannot be suppressed,
argues Smith, and it now increasingly explodes through the tunnel.
Indeed, considerable numbers of scientists are themselves now asking
metaphysical questions.

So also the human capacity to find life more fulfilling through helping
others does not go away. Members of A.A. understand that as they help
other alcoholics, they also help themselves. This principle is clear in the
purpose statement of A.A.: "Our primary purpose is to stay sober and
help other alcoholics to achieve sobriety" (A.A. 1939/2001, 100). There
is a relevant aphorism: "If you help someone up the hill, you get closer
yourself." There is a deep sense of purpose in such a role, and a powerful new self-identity as a "wounded healer," one who assists others
from one's reservoir of firsthand knowledge.

A.A. is a success story for those who resonate with it and apply it
over the years. Does the ontological generality upon which the twelve
steps are constructed point back to some outmoded image of human
fulfillment that we can now jettison as archaic and useless? Or does it
point us forward to images of human fulfillment and increased health
in which the dual axes of spirituality and helping others combine in a
synergy of energy and practice that can serve all humanity well, reduce
health-care needs and costs, and contribute greatly to human progress

in a time when care of the self needs a new foundation? I believe that Sir John would view this as a prototype for the future, as do I. Nothing is proven here, but the fulfillment that comes through abiding in the ontological generality suggests that it is the ontological reality into which we have evolved spiritually.

EVIDENCE 3: THE PLAUSIBILITY OF AN UNDERLYING GROUND OF BEING OR ULTIMATE REALITY THAT CONSTANTLY CREATES AND SUSTAINS THE UNIVERSE

The evidence for Unlimited Love as Ultimate Reality is significant from what a great majority of randomly selected Americans self-report as their essential spiritual experience, although this could be further developed with a worldwide survey. In addition, individuals who abide in the double love commandment of love of God and of neighbor as self appear to be relatively happier and healthier, suggesting that human beings flourish in this triadic nexus of love. These evidences are now rather well demonstrated.

The third line of evidence relies on the physical sciences. It is the claim that a Ground of Being—an Infinite Mind with Creative Love Energy—gives rise to the big bang, the laws of the universe, and fine tuning to support the emergence of carbon-based life, and also constitutes the prime energy sustaining the universe. This Ground of Being is a substrate underlying visible reality, and a supporting matrix of prime thought energy. This Ground of Being is plausible unless one wishes to assert (a) that something can come from nothing, or (b) that there are some billions of parallel universes (or perhaps that one universe precedes another in an infinite regress), such that probability would allow that one of them would be like ours in being generative of life. The alternatives may be plausible, but they are highly speculative and seem to be dreamed up to make it arguable that there is no Ground of Being or Ultimate Reality.

A lot of people have trouble with the word "God" because it has so many anthropomorphic associations with human images. To speak of Ultimate Reality or Ground of Being is to get as far from a white-haired old man sitting on a throne as possible. This is why such language became current in twentieth-century theology. By Ultimate Reality we mean the essence of all being, that in which we and all creation "live and move and have our being" even if we are unaware of it. This Ground of Being would have the qualities of Prime Energy, Prime Mind, Prime Creativity, and Prime Love.

THE BIG BANG

There was never nothing, because Prime Energy is timeless and eternal, although it became manifest in the form of our universe suddenly in an explosion roughly 13.7 billion years ago from a single point, like a massive fireball giving rise to all that is. The energy from this event still vibrates as the cosmic background radiation of the universe. It was all so hot that when it cooled in a thousandth of a second to 100 billion degrees centigrade, elementary particles such as electrons, protons, and neutrons could form. And a minute or two later the temperature dropped a little further, allowing neutrons and protons to combine to form atomic nuclei, initially hydrogen and helium. Expansion and cooling continued, and after 700,000 years at the temperature of the sun simple atoms began to form. Galaxies and the first stars formed. Eventually there came the generative planet earth.

Without needing to offer details, the point here is that no one can simply assert that our universe is eternal, and that therefore, because it always existed, we need not even think about some First Cause.

Not everyone had an easy time accepting the big bang theory, including the English astronomer and mathematician Fred Hoyle (1915–2001), who mockingly named the theory "big bang" in 1950 on a BBC radio program, *The Nature of the Universe*. Hoyle was a committed atheist humanist who rejected the big bang nearly to the end of his life long after it was considered proven. Hoyle invented his "steady state theory" in part because the big bang theory of a beginning to the universe points to there being a Beginner, and Hoyle would have none of that. His idea of a steady-state "creation field" did have some interesting theoretical underpinnings, however, and for some time it was an equal contender with the big bang theory. Many scientists accepted it for a decade or more because it removed any need for a Beginner. Yet Hoyle, it is said, did eventually come to assert a "superintelligence" acting at the quantum level (see Davies 1992, 229), and he understood all of reality as constantly formed and shaped at the quantum level by something that is at least very godlike.

Of course, we know now that the universe did have its beginning in a big bang, a mega-creation event. Everything in our observable universe is the product of a huge explosion that occurred by the most exact calculations 13.7 billion years ago. Since then space itself has been stretching, and thus the universe is expanding, like the surface of a balloon. At the zero point those 13.7 billion years ago, all the energy we see in our universe was compressed into a single infinitesimally small point (the "singularity") that blew apart, like a primeval atom.

When this idea of what came to be known as the big bang was first presented by the Belgian astrophysicist and priest Georges Lemaitre (1894–1966) in 1927 in the *Annals of the Scientific Society of Brussels* as the "hypothesis of the primeval atom," many scientists were skeptical because the evidence had not fully accumulated, and because it seemed too much like a religious idea. Lemaitre did actually propose the expansion of the universe, and he did derive what is known as the Hubble constant. Two years later, however, the American astronomer Edwin Hubble came out with his idea that the galaxies were receding from each other and the universe was then expanding. But it was Lemaitre, a Cambridge product teaching at the Catholic University of Louvain, with whom the big bang originated. Lemaitre had worked with Arthur Eddington at Cambridge, and then spent time at the Harvard College Observatory with Harlow Shapley before returning to Belgium in 1925. The title of his 1927 paper was, "A Homogeneous Universe of Constant Mass and Growing Radius Accounting for the Radical Velocity of

Extragalactic Nebulae." He actually derived Hubble's law in this paper and made the first observational estimate of Hubble's constant, although Hubble did go further than Lemaitre regarding the numerical value of the constant based on Hubble's additional data and observations. The Lemaitre paper was published in a journal not widely read by astronomers but was republished in 1931. In 1931 Lemaitre returned to MIT to defend his doctoral thesis and then took up his position at Louvain.

Einstein and Lemaitre met in 1927 in Brussels, and Einstein took no exception to the mathematics in the 1927 paper but was not convinced of the expanding universe. The two met on several other occasions, including the Solvoy Conference and in 1935 at Princeton. They traveled together to California for a series of seminars. On one of these California occasions, Einstein stood up, applauded Lemaitre's lecture, and commented that it was the most beautiful and satisfactory explanation of creation he had ever heard. Einstein proposed him for the Francqui Prize, the highest Belgian scientific distinction. In 1953 he received the first Eddington Medal awarded by the Royal Astronomical Society. He died in 1966 just after the discovery of the Cosmic Microwave Background Radiation (CMBR), a faint light emanating from the big bang explosion that gave further evidence to his discovery. Later, in the 1980s, the theory of an initial period of inflation was added to the big bang theory, which by then was universally endorsed. The big bang almost certainly occurred with a brief "inflationary" moment of huge increase in size. So we have the Standard Big Bang Model (SBBM) in which there was nothing before the big bang, including no space and no time, no matter and no energy as we know it.

Now, as soon as we speak of a beginning point prior to physical reality, energy-as-we-know-it, time and space, we must ask from whence these things come. Well, if something cannot come from nothing, we can postulate a universal and original Prime Energy that is the Divine Energy itself as manifested in the big bang, and compressed in some small portion into that "primeval atom" or "cosmic egg." Everything that exists in our universe is grounded in this Prime Energy, which constitutes the external form of God. All of reality in its manifestations as energy and matter is Prime Energy, and its associated manifestations of Mind, Creativity, and Love.

THE LAWS OF THE UNIVERSE AND FINE TUNING

The idea that the fundamental laws of the physical universe themselves provide evidence of Prime Mind contrasts with the idea that the violations of these laws in the forms of the miraculous is of interest. The laws themselves are already miraculous enough.

The atheistic physicist might return the favor and say that even if there is a Prime Mind, who created this Creator? Well, the answer is that the Creator has no beginning, transcends time and space, and is eternal in a way that things caused by it are not. Or the atheist might suggest that the laws of the universe "may be eternal, or they too may have come into existence, again by some yet unknown but possibly purely physical process" (Krauss 2012, 142). How could such intelligent laws not come from an intelligent source?

Consider the energy constants, which are the four fundamental forces: gravitational, electromagnetic, strong nuclear, and weak nuclear. We find the gravitational attraction constant, the weak force coupling constant, and the strong nuclear force coupling constant. There are constants of electron or proton unit charge, and there are constants such as Boltzmann's constant, the cosmological constant, and so forth. Our universe clearly seems governed, ruled, and organized by a whole set of constants and universal laws of physics and mathematics. It is all quite elegant and so improbable that this could all just exist without an intelligent cause Prime Mind and some sort of Prime Force. Why would one think that the rules and constants of a Monopoly game were not dreamed up and manifested by a creative mind? To suppose otherwise would be bizarre.

How can we not assume that these laws are a manifestation of Prime Mind (i.e., of that which composed them and made them manifest)? Where else might they come from? Stephen Hawking, in *A Brief History of Time*, suggests that if we were to really understand and dig deeper into physics to discover what he calls "a complete theory," "then we would know the mind of God." Of course, Hawking remains agnostic, and he is speaking more metaphorically than literally. But even he seems to acknowledge that something lies behind these laws.

These things came into being with the big bang and so they must have existed in the Prime Mind. Brahman (Hinduism) and *Ein-Sof* (Kabbalists) are names for an ultimate and eternal reality of Godhead that is mathematical and ingenious and bursting forth in creativity. These laws and the like have to be conceptualized, manifested, and sustained by something. They are manifestations of Prime Mind, of what Neoplatonism referred to as *Deus Absconditus*, a hidden God. Over the centuries of modern physics many of the great minds, from Newton to Einstein, have assumed this to be so.

Very closely associated with the argument that the existence of the laws of the universe implies a Godhead is the fact that these laws seem to be fine tuned for carbon-based life. Had the constants been just slightly different, there could be no life. Had the big bang explosion exploded just slightly more intensely, matter would have been blown apart too fast for stars and planets to form. Had the gravitational force been infinitesimally smaller or larger, stars could not have come into existence.

Those convinced that our universe is just one of billions of universes would say that at least one of them would have the right conditions for life to emerge, and we are the beneficiaries of this. But even in this debate there is the considerable consensus that something needs to be explained either on the basis of God or multiple universes. Maybe there are those trillions of universes existing simultaneously or in sequence, but their hypothetical existence is really an artifact of a theorizing that would only arise on the assumption that there is no "superintellect."

Many contemporary physicists are driven by science toward this idea of an Ultimate Reality. The laws of physics and math, the constants of the universe, seem to be so carefully thought out and organized. This organizational elegance of the universe is so improbable that the idea that all of these background conditions for life arose by pure chance is almost infinitely unlikely.

Devoted atheists and agnostics attempt to explain the improbability of our anthropic universe by appeal to superextensivity. It is just "said" that there are trillions of universes, and that by chance one of these would be anthropic. However, no one has shown that any of these other universes exist, much less seen them. Ideas for a naturalistic explanation of our anthropic universe, in contrast to a transcendent one (Ground of Being before space and time), are all highly theoretical and have no clear empirical basis (e.g., the Everett-DeWitt quantum "many worlds" hypothesis, the string theory landscape). The argument for a transcendent cause (a transcendent intelligent agent) appears relatively more likely. Thus do considerable numbers of physicists lean toward the high plausibility of a Prime Mind that has set up our universe precisely as it is.

THE ENERGY SUSTAINING THE UNIVERSE

Moving beyond ordinary experience and perception to the subatomic world, science shows us a whole new world. We find particles darting into and out of existence in much less than a billionth of a second. A gluon (a particle within a nucleus that "glues" things together) lasts about that long between its creation and annihilation. A neutron lasts about fifteen minutes, far and away the most long-lived of particles. In the quantum world an electron can appear like a little particle or like a wavelike cloud that moves at speeds as high as the speed of light. The quantum world deals with so many unobservable quantities (wave functions) and leaves the boundary between perception and quantum reality vague. Probabilities replace certainties, quantum objects can be at more than one place at the same time (the wave property of electrons) and they can jump without going through intervening space (quantum jump), and things can't be seen directly because almost as soon as they exist they decay out of existence in quick annihilations. One might say that the quantum world is more like a Jackson Pollock painting viewed while on peyote than the stable image of a British landscape painting. The latter represents the Newtonian world, but Pollock's world is the substrate.

The universe does not seem solid or material, but immaterial and unstable. It seems real but unreal. Matter turns out to be fairly bizarre stuff at the quantum level. It seems almost immaterial. Nothing lasts as things flicker in and out of being. Atoms turn out to be mostly empty space and waves. The proton in the nucleus of a hydrogen atom is like a baseball sitting on the pitcher's mound in a stadium, with an electron about the size of a fly sitting out in the upper decks but also being a wave. Matter turns out to be rather ephemeral, with bits and pieces that are not like grains so much as miniclouds of charge and forces.

At this level, Ground of Being is understood as the integrated stabilizing and sustaining underpinning of the strange world at the quantum level.

Thus it can almost be said that Ultimate Reality is a complex of energies and forces that seem immaterial in quality. Clouds and wave functions seem to be pervasive.

Such a quantum world is of great interest to philosophical and spiritual thinkers, for it would seem that there is room for the constant sustaining Prime Energy, Mind, and Creativity. Nothing is proven, of course. But quantum physics makes God more plausible by undermining the materialistic philosophy that is often appealed to in ruling out God.

CONCLUSIONS: TAKING THE IDEA OF GOD AS UNLIMITED LOVE INTO THE SUBSTRATE AND HIDDEN MARTIX OF THE UNIVERSE

The existence of God as Prime Mind and Energy is plausible. But what about Creative Love? I would rather ask how God could be considered seriously without including Prime Creativity rooted in Love.

In the *I Ching*—the classic book of Chinese philosophy and divination—the first of sixty-four hexagrams is called simply "Ch'ien"—the creative. As the *I Ching* explains, creativity is the force that stands for the primal power of the universe. The now deceased Harvard theologian Gordon Kaufman has suggested that we think of God not as a creator but as creativity itself. In his book, *In the Beginning ... Creativity*,

Kaufman begins with a paraphrase of the opening verses of the Gospel of John in the Bible: "In the beginning was creativity, and the creativity was with God, and the creativity was God. All things came into being through the mystery of creativity." Kauffman suggests that this view of creativity "preserves the notion of God as the ultimate mystery of things, a mystery that we have not been able to penetrate or dissolve—and likely never will." And, says Kauffman, though we can construct a picture of creativity in the universe—from the big bang, through the evolution of star systems and planets, to the evolution of the extraordinary diversity of life on earth, to the emergence of human beings and history—we still cannot quite picture or embrace the ultimate point of reference that we call God.

Creativity is generally closely linked to love. Surely much of human creativity is driven or shaped by the dynamic of love, whether in the family, between friends, in the neighborhood, at work, or for all humanity. From Bach's pieces for his daughter Anna, to Eric Clapton's haunting tune in memory of his son, so many wonderful musical compositions are dedicated to a particular loved one. Haven't most of us written a letter with special creative care to convey love to one of the kids, or spouse, or an old friend fallen ill? Creativity is frequently inspired by the sense that the life of another is at least as meaningful as one's own, or more so. *This is the dynamic of love that shapes creativity* in preparation, incubation, and illumination.

When I see that mother who has great love for a young child knitting an intricately designed blanket, when I encounter an artist or composer creating something wonderful in dedication to a loved one, or when I see someone thinking with such care about what sort of gift to give to a friend, I see a connection between love and creativity that is both an everyday constant and an absolutely beautiful human tendency. So often we create "simple gifts" of handcraft that don't measure up to museum standards but are heartfelt expressions of creativity as an expression of love.

The birth home of inventor Thomas Edison is about two hours west of Cleveland, in the town of Milan, Ohio. Edison was in large part inspired by a profound love of all humanity. "The dove is my emblem. . . . I want to serve and advance human life, not destroy it. . . . I am proud of the fact that I have never invented weapons to kill," he stated. Moreover, "I never perfected an invention that I did not think about in terms of the service it might give others. . . . I find out what the world needs, then I proceed to invent." In his adult heyday as the "wizard" of Menlo Park, New Jersey, he affirmed, "Personally, I enjoy working about eighteen hours a day. Besides the short catnaps I take each day, I average about four to five hours of sleep per night." So how did this man—who only had three months of formal schooling as a child before his teacher determined that he was "brain scrambled" and did not belong in school, who had to be home-taught by his deeply benevolent mother, who at age fourteen became totally deaf in his left ear and 80 percent deaf in his right, and who would for a while be a young beggar on the streets of New York—define his creativity? "Genius," he stated, "is 1 percent inspiration and 99 percent perspiration. Accordingly, a 'genius' is often merely a talented person who has done all of his or her homework." His noble purpose to enhance humankind allowed him to endure failure after failure, including the more than ten thousand failed attempts at finding a filament for an electric lightbulb. "If I find ten thousand ways something won't work, I haven't failed." The outcome? In addition to the lightbulb, the electric motor, the phonograph, the alkaline battery, and countless other devices that we all today take for granted, Edison accumulated 1,368 patents in all! It is good that his mother, a Presbyterian, taught him that his gifts were for benevolent purposes. This seems to have gotten him over some rough patches.

The Hebrew Bible begins with a poem about creativity. "In the beginning, God created the heavens and the earth." It started with darkness and chaos, with turbulence and unharmoniousness, upon which descends the cherishing creative energy of a loving God. Creativity envisions and hopes, broods over the formless void and the unpromising depths, and in ceaseless movement begins. The Sculptor of the universe, with patient fostering love, shapes the universe from unformed substance. Creation comes into being through hovering and brooding

love, bit by bit, unhurried, as Creative Spirit draws forth latent possibilities for beauty. Even when nothing seems to be happening there is the
forming of a vision from which beauty will appear. And so Eternal Love
and Eternal Creativity are one. The God of the Hebrew Bible is the first
artist. And if we wish to believe it, we are formed in the image of a loving
artist—or if you prefer, we envision a God whose attributes are those we
most admire in ourselves. (Indeed, we are made "in the image of God"
because we share in the Divine Mind and Creative Love.)

Prime Mind was motivated to create through a big bang, to sustain
the incredible constancy of the universe through its many laws, and by
forming it. The classic answer put forward by the perennial philosophy
is that the Godhead (Brahman) sought to increase the potential for love
and creativity through giving a grain of the larger Divine Mind to each
of us. I would suggest that God is a perpetual, self-generating energy,
the first cause and primal source of all that exists and the sustaining
matrix underlying all things. The universe is a manifestation of this
ultimate source of energy, and it is thus God's outer form. This dynamic
God of energy and intelligence has also the essence of personality. The
essence of this God is Love itself. Bhakti yoga in Hinduism, Sufi Islam,
Tantric Buddhism, Hasidic Judaism, Catholic Franciscans, and all mystical people pursue and cherish the devotional love of God.

God's intrinsic desire in creating and sustaining the universe is to
feel joy. Hence, the God of Love projected God's whole nature into
creation. The universe was not created out of nothing but out of God's
Love, Intelligence, and Energy, and this Ultimate Reality permeates
every part of the universe.

The effort to grasp the origin, development, and structure of the
universe is of course not simply a modern scientific endeavor. The
Babylonians, Egyptians, and virtually every ancient culture has their
cosmological myths, with some of them more and some of them less
consistent with scientific cosmology. But from either approach, the
problem of the cosmos with regard to origin, development, structure,
and sustainability is perhaps the single most engaging area of intellectual inquiry, particularly at the interface of science and religions. As

science has progressed from Newton to Einstein to quantum theory and astrophysics, all useful theophilosophy has rightly followed along and incorporated these exciting developments into its core; as a result, theologies become progressive and engaging. Otherwise, they more or less move to the margins of intellectual significance. Indeed, as the scientific instruments of modern technology are employed, the unaided cosmologies of the ancients often turn out to be wrong or extremely limited. When religions engage in conceptual evolution consistent with modern cosmological science, we have what John Templeton referred to as "spiritual progress." Premodern cosmologies need to be updated because things are not what they seem to the naked eye.

It seems plausible to assert that the universe depends on Unlimited Love. This idea of ultimate dependence captures the metaphor of God as the Ground of Being, and equally, of Ultimate Reality. So do theophilosophers say that the universe depends on God, who precedes the universe in a spaceless and timeless prime force and energy that must include creativity and some motive to create, the most sensible motive of which is love. This ultimately comprises this Ultimate Reality, and so in this sense, Sir John could ask, is God the only reality?

Unlimited Love as Ultimate Reality? This phrase means that the very eternal disposition of this Ground of Being is love, a desire to cherish and nurture other free and spiritual beings. Unlimited Love is God's very being, even the very essence of the force and energy of God from which the entire universe came into being and remains in being. Unlimited Love is the eternal disposition or fundamental attunement of God who supplies the motive underlying divine creativity. Love is the dynamism behind the big bang, the constants, the forces, and all that is. But inanimate objects and various forces are merely the building blocks from which our anthropic universe allows the evolution of a life form that has the qualities of spirit, self-consciousness, mind, and cocreativity in potential relationship with God and in the divine image. This perspective does not sound all that crazy once one reads the literature of the purely naturalistic physicists.

This work is ultimately about progress in spirituality and religion, and about a common quest for joy. We suffer from spiritual or internal, and scientific or external, ignorance. This suffering is killing our species. Why do we exist? What is the meaning of life? Why be moral? Does God exist, and how much do we really understand about this? These are the internal questions. What is the cause of the universe? What sustains it? What is its future? What are the most fundamental particles? These are external questions. Through philosophy and religion, the human species has tried to answer the internal questions, and through the sciences it has tried to answer the external ones. What religion and science have in common is a desire to find truth and overcome ignorance.

Sir John Templeton saw this common desire, and he realized how extensive and deep our ignorance is in both religion and science. He believed that this ignorance can be best overcome when the two domains of knowledge move forward in consonance, centered on the purest and most rigorous unhampered scientific methods—and on humility-in-theology as we realize how little we know about perennial spiritual truths such as "God is love," but do not jettison them. He sought integration between science and religion, but he completely rejected religion interfering with the process of scientific investigation. Hence, the John Templeton Foundation funds high-level science and then awaits a final report to see what has been learned and its possible bearing on religion and spiritual progress.

Sir John hoped for real spiritual and religious progress with the help of science. His dream—and the world needs dreamers—was for a time when all the different methods of seeking truth would in perfect methodological integrity and in harmonious mutual respect converge not superficially but deeply on the real secret of the universe—that Unlimited Love is Ultimate Reality—and from this point of convergence hostility and discord would end, traditional difference and hatreds would be set aside, and a true family of humanity would be able to flourish on this earth and in eternity in true spiritual progress.

One of the best few things that ever happened to me in my adult life was getting on that train with Sir John. In writing this book, I hope that those who come long after me will buy a ticket at the same station, because the best thing any of us can do here on earth is learn about and practice the one thing, Love, that means the most in heaven. I do believe that when people experience Unlimited Love in the very depths of their being, they are in fact experiencing Ultimate Reality, and that someday science will say it's so.

Sir John was excited about the future of science and spirituality. He wrote these words of immense optimism:

> We are perched on the frontiers of future knowledge. Even though we stand upon the enormous mountain of information collected over the last five centuries of scientific progress, we have only fleeting glimpses of the future. To a large extent, the future lies before us like a vast wilderness of unexplored reality. The God who created and sustains His evolving universe through eons of progress and development has not placed our generation at the tag end of the creative process. He has placed us at a new beginning. We are here for the future. Our role is crucial. As human beings we are endowed with mind and spirit. We can think, imagine, and dream. We can search for future trends through the rich diversity of human thought. God permits us in some ways to be co-creators with Him in His continuing act of creation. (1990/2006, 209)

Sir John believed that the more we learn, the more we will realize that "every person's concept of God is too small" (1990/2006, 210).

Sir John asked in his letter to me, "Could love be older than the Big Bang?" Yes, it could be. He would not have expected proof, but plausibility. I do believe that he was correct in writing of "the Great Face Behind," and in asserting, "The manifold scientific discoveries of the late twentieth century cause the visible and tangible to appear less and less real and to point to a great reality in the ongoing and accelerating creative process within the enormity of the vast unseen" (Templeton and Herrmann 1994, 9). I see no reason that Unlimited Love could not be older than the big bang, and find it hard to imagine that some such emotive force does not in fact underlie and motivate the creation and sustaining of all that exists. As we peel back the levels of the cosmic onion to more and more ultimate levels, we may find that the perennial philosophy is, after all, correct—that in the chaos of the lowest ultimate order of energy and matter where particles are boiling up into momentary being there lies an Ultimate Reality of Unlimited Love in which we live and move and have our being. It would be plausible that at the motivational level, Unlimited Love as an aspect of the Divine is in the very energy and material of the universe, mediating a dynamic flux of constant creativity, constant transition, constant becoming, and constant vanishing. As Sir John wrote, "Perhaps in the end the only reality is God." (Templeton and Herrmann 1994, 163)

As Sir John also wrote, "Perhaps, in the end, Reality will appear far deeper and more profound than our limited human abilities can ever hope to comprehend" (Templeton and Herrmann 1994, 25). But in the meanwhile, "It should be our finest goal to know God by every avenue open to us. Surely the best and most profound studies in all of history will come when performed humbly with the expectation of more fully knowing the Creator God and his purposes for his creatures." (Templeton and Herrmann 1994, 170)

Sir John Templeton was an ontological gospeller. He saw in this universe from the stars in heaven to the smallest particles the work of Unlimited Love, and he celebrated this with creativity and love. He drew on the perennial ontology of the Gospel of John repeatedly, and he surmised that someday, because truth is One, science and mysticism would see eye to eye on this.

PART 3

Statements from Family Members,
Associates, and Scholarly Friends

Introduction to Part 3

I N JUNE 2012 I wrote to a number of family members and associates of John Templeton as follows:

> I am writing with a manageable and very important request in the aftermath of the spectacular late June 2012 celebrations of Sir John's 100th birthday, the 40th anniversary of the Templeton Prize, and the 25th anniversary of the founding of the Templeton Foundation. The events were fabulous, including a Laws of Life day in the town square of Winchester, TN, attended by an estimated 2000. On Monday, June 25, we had a full-day conference on "Spiritual Progress and Human Flourishing" with wonderful speakers reflecting on what Sir John meant by "spiritual progress" and the extent to which such progress has been made in their topical areas ranging from Forgiveness and Unlimited Love to Cosmology and Awe. Truly, this was a week I shall not forget. The request is one that I have selected you for based on your having known Sir John and on your familiarity with his creative and loving mind. I simply invite you to offer some several pages either on (1) what you think Sir John meant by "Unlimited Love as Ultimate Reality," or on (2) how you experienced him as a person who tried to live a life of love. Some of you may wish to address both of these if you are able to do so. I am not intending to edit these statements, for I prefer to let them stand entirely as you have written them in order to form a clear historical archive.

I received a fascinating set of responses, which I present on the following pages.

Chapter 10

What Do You Think Sir John Meant by "Unlimited Love as Ultimate Reality"?

PROFESSOR OWEN GINGERICH, HARVARD
UNIVERSITY COSMOLOGIST

Essay on John Templeton and Unlimited Love

Two and a half decades ago, when the John Templeton Foundation was just being established, I had been working intensively with Robert Herrmann, the executive director of the American Scientific Affiliation, hoping to produce a blockbuster television series. This provides a significant clue for understanding how I entered Sir John Templeton's circle, for at the same time Bob Herrmann had become acquainted with Sir John and was helping him recruit a board of advisors for his proposed foundation.

First, a word about the television episodes that never came to be. In 1980 Carl Sagan's *Cosmos* series became the most widely watched PBS program up to that time. In addition to a fascinating picture of astronomy past and present, the *Cosmos* series purveyed an undisguised materialistic spin. As the guiding light of an organization whose members took both the Bible and science seriously, Bob Herrmann hoped to create an alternative TV series, one that honored the Judeo-Christian background out of which modern science arose. I agreed to write and moderate the programs, and we worked extensively with Geoff Haines-Stiles, who had been one of the producers of the *Cosmos* series.

With Bob Herrmann as an intermediary, I received an invitation from Sir John to join his board of advisors, and he was delighted when I finally agreed. Sir John gradually established a fascinating board of advisors, with some of my old friends, others I knew only by reputation, and still others who represented areas previously outside my own circle. So many interesting people agreed to serve that attendance at the meetings was very high, since the members enjoyed getting acquainted with individuals with very different professions and accomplishments.

Typically Sir John invited me to sit with him at one of the advisory committee luncheons. He had an unusually fertile mind, very alert and curious about a wide range of topics. He always asked me what was new in astronomy, and I inevitably tried to bring some hot new tidbit to satisfy his desire to keep up with the latest science. At one lunch he told me, "There is someone here I want you to meet. He is thinking about giving a prize."

And thus it was that I met Peter Gruber, who had been one of Sir John's most successful investment managers because he had an excellent background in overseas markets. Peter recounted his own extraordinary life journey, being a young refugee from Nazi Europe, having a Jesuit education in India, starting college in Australia, then beginning at the bottom on Wall Street, going to South America, and becoming a global investor in emerging markets. It was in India where a priest had shared his telescope, and Peter got his first taste of the stars. Now he proposed to found a major prize in cosmology, especially because there was no Nobel Prize in astronomy.

The advice I gave Peter Gruber was partly right and in one critical respect dead wrong. I said he should make the prize broad enough to cover important new contributions in astrophysical astronomy that might not be narrowly defined as cosmology (which he accepted), and I also said that there were probably not enough developments in what would be considered cosmology to find a high-quality recipient every year (and there my crystal ball was very cloudy). Two years later Peter and his coconspirator Patricia set up the cosmology prize (and I am happy to say I am still an advisor to the Gruber Foundation for their annual, highly regarded prize).

Meanwhile, even before I joined Sir John's advisory board, I was on a small group of advisors for the Center of Theological Inquiry in Princeton, an institution that was from time to time recipient of Sir John's benefaction. One day when I was there for an advisors' meeting, I saw that the distinguished physicist Freeman Dyson was also in the building. When a friend who had, along with me, been a founding member of the Mennonite Congregation of Boston but who was then living in Princeton, arrived to pick me up, I noticed that he stopped for a conversation with Dyson. Later I asked him how he knew the Princeton physicist. "Oh, he comes to our church," was the reply. It had never occurred to me that Dyson was a churchgoer, so I asked him if he would consider joining the group of Templeton advisors. Sir John was tickled by this prospect and promptly invited him to become a member. Dyson often sat with me at the advisors' meetings, and we frequently passed notes back and forth commenting on the proceedings. He seldom if ever volunteered advice, but responded wisely whenever he was specifically called upon. Eventually Dyson received the prestigious Templeton Prize.

Of course Sir John was aware that Bob Herrmann and I were diligently seeking funds for our television project. He was very cool to the idea on the grounds that opinion leaders never watched television, but got their ideas from reading. He was completely impervious to the argument that the leaders of the future were young people addicted to television. We were looking for a million dollars, which we hoped to leverage as a matching fund from other agencies that would then take our proposal seriously. Our budget was modest compared to Sagan's kitty of over $10 million. Nevertheless, I think Sir John was troubled by the size of our request, especially when someone offered to make a film about him for a few tens of thousands of dollars. I explained that one could make a modest museum exhibition for $50,000, but a "blockbuster" room at the National Air and Space Museum typically cost over $2 million. And I pointed out that his church science lectures with travel and honorarium expenses for the lecturers were costing about $50 per person-hour in the audience, whereas we could offer a quality audience for only a few cents per person-hour. Sir John was unmoved despite the fact that a majority of his trustees favored the project. At least some of the ideas developed for the never-to-be TV series resurfaced in my Harvard lecture series and book, *God's Universe*.

During the 1990s, when Sir John was in his eighties, he was continually generating new ideas, and this included a number of books, some of which he wrote himself or with collaborators, and some that he commissioned with multiple contributors. He was always very future-minded, having little patience with history. Consequently I couldn't help but head one of the essays I wrote for him with a quotation from Kierkegaard: "We live forward, but to understand we look back." Whether he ever noticed it, I know not!

Among other book projects, Sir John envisioned one on the near future, the prospects for the next four decades in a variety of fields. He invited me to take on the challenging task of describing physical science in the 2030s. I found it irresistible to see how past futurologists had fared in their predictions, and in a vast barn full of secondhand books in New Hampshire I found a large cache of such material from the 1920s and '30s. Disposable paper dresses, driverless cars, and inoculations to prevent tooth decay were predicted by 1980. The ability to control rain, fog, and the climate was envisioned by 2018. Then there were a few right on target, including a 1902 prediction that by 1950 there would be heavier–than-air flying machines.

As for the future, I mentioned finding planets around other stars, the Higgs boson, controlled nuclear fusion, computer modeling of future trends such as global warming, and earthquake predictions. Finally I concluded with a quotation from the seventeenth-century astronomer Johannes Kepler: "I consider it a right, yes, a duty to search in a cautious manner for the number, sizes, and weights, the norms of everything He has created. . . . For these secrets are not the kind whose research should be forbidden; rather, they are set before our eyes like a mirror so that by examining them we observe to some extent the goodness and wisdom of the Creator."

Sir John was not merely in favor of that—he set up his foundation to include funding for such endeavors. He was fascinated by what astronomers call fine tuning, which I wrote about in his 1994 book, *Evidence of Purpose*. I mentioned the remarkable properties of the carbon nucleus, predicted by Fred Hoyle, which leads to the high abundance of carbon atoms in the universe, and which in turn makes complex life possible. The specific resonance levels in the carbon nucleus and those in the oxygen nucleus could not differ by more than a few percent without suppressing the high abundance of carbon. Fred Hoyle, who was essentially a skeptic, nevertheless famously wrote, "Would you not say to yourself, 'Some supercalculating intellect must have designed the properties of the carbon atom, otherwise the chance of my finding such an atom through the blind forces of nature would be utterly minuscule.' A common sense interpretation of the facts suggests that a superintellect has monkeyed with physics, as well as with chemistry and biology, and that there are no blind forces worth speaking about in nature. The numbers one calculates from the facts seem to me so overwhelming as to put this conclusion almost beyond question."

By a superintelligence we can no doubt envision a brilliant wizened scientist twiddling knobs on a vast control board in his laboratory. To Sir John this was interesting but far from making the point about the deep, fundamental nature of the universe, a universe made for us and caring for us. Sir John's was a trinitarian cosmology of sorts, not in the traditional sense, but a triangle with God at one corner, Ultimate Reality at another, and Unlimited Love at the third vertex. Linking it all together was 1 John 4:8: "He that loveth not knoweth not God, for God is Love" (King James Version), or "Whoever does not love does not know God, because God is love" (New International Version).

With his disdain of history, Sir John might have been very surprised to learn that more than two millennia ago Aristotle was already partway there. In his *Metaphysics*, Aristotle inquired into eternity and final causes. What moves without being moved? Aristotle asked. It must be the eternal heavens, with unceasing movement, he responds. But why? It must be desire for the good, and the final cause is therefore love. In the climax to this passage, Aristotle writes, "If then, God is always in that good state in which we sometimes are, this compels our wonder; and if in a better state this compels it yet more. And God *is* in a better state. And life also belongs to God; for the actuality of thought is life, and God is that actuality; and God's self-dependent actuality is life most good and eternal. We say therefore that God is a living being, eternal, most good, so that life and duration continuous and eternal belong to God; for this is God" (*Metaphysics* 12.7).

Aristotle's term *metaphysics* is, literally, "beyond physics." The study of physics, both by theorists with their pencils, paper, and calculators, and by experimentalists with their giant accelerators, is knocking at heaven's door, an approach to ultimate reality. Sir John recognized this, and included funding for such "big questions" in his foundation. But there must be more, the "beyond physics," the questions of the origins of laws of nature, why the universe is comprehensible, why there is something rather than nothing. This is the zone of spiritual progress, a key to his "so eager to learn" philosophy. This is why it is not mere reality, but *ultimate* reality.

But in this conceptual scheme, should there not be a clearer link between a Creator God and his Created Creatures, endowed with creativity, conscience, and self-consciousness? How can we understand the existence of the exquisite design that makes possible not only this physical world but also thinking beings—thinking beings who wrestle with understanding not only the matter and energies composing the universe but its purposes and relationships? Sir John was a wrestler, with a vision of spiritual progress. Surely the link between ultimate reality and God is a manifestation of Unlimited Love, the glue that holds this remarkable vision together.

GEORGE ELLIS, PROFESSOR OF PHYSICS,
UNIVERSITY OF CAPE TOWN

Unlimited Love as Ultimate Reality?

Sir John was an original and creative thinker. He saw clearly, and was deeply impressed by, the huge strides that science had made in the past three centuries through its impersonal analysis of causation, but he was also deeply aware of the importance of meaning and values, and he wanted to somehow link them together. Furthermore a center of his thought was the idea that ultimate purpose was about unlimited love. He wanted to see these ideas united together. This led to his idea that somehow agape love has its place in the grain of the universe (Ellis 1993).

This fitted in well with my own research program, developed primarily through my interaction with the excellent Vatican Observatory/ CTNS program run by Bob Russell, George Coyne, Bill Stoeger, and Nancey Murphy. I was heavily influenced by them and by others such as Arthur Peacocke, Ian Barbour, and Phil Clayton, who were also part of that program. My views are set out in three writings that present a view broadly concomitant with that of Sir John (Ellis 1993; Murphy and Ellis 1996; Ellis 2011). Through this work I became involved with the Templeton Foundation from 1996 on, and benefited much from the many opportunities this brought my way. I used to see Sir John regularly at foundation meetings until his death, and am thankful to him for his vision that made that all happen. He has played a profound role in promoting the recent development of the understanding of the relation between science and religion.

The basic scientific view is set out in the bifurcating hierarchy of scale and causation presented in Table 1, showing the natural sciences hierarchy (left) and the human sciences hierarchy (right). The basic three levels are common to both the natural sciences and the human sciences branches: all matter is made out of the same basic stuff. This is one of the extraordinary features of physical reality: both inanimate matter and life emerge from the same substrate of inanimate atoms.

Table 1

THE BIFURCATING SCIENCES HIERARCHY

Level 8	Cosmology	Sociology/Economics/Politics
Level 7	Astrophysics	Psychology
Level 6	Planetary/Earth science	Physiology
Level 5	Geology	Cell biology
Level 4	Material science/Mineralogy	Biochemistry
Level 3	Chemistry	
Level 2	Atomic physics	
Level 1	Particle physics	

Source: Murphy and Ellis 1996.

A key feature that enables true complexity is that, in contrast to reductionist understanding, causation goes both bottom up and top down in this hierarchy, enabling interlevel feedback loops (Ellis 2012b; Ellis et al 2012). It is not true that higher-level behavior can be reduced to "nothing but" lower-level interactions.

Given this picture, does physics lend plausibility to the idea that the Ground of Being might be this Ultimate Reality? Table 1 does not consider such issues: we have to ask what higher and lower levels there may be. This is indicated in Table 2.

First, particle physics is based on the specifics of the laws of physics: their nature and the values of the fundamental constants that determine the relative strengths of the various forces (Level 0). The key question is why these forces are as they are, particularly because it is clear that if these values were markedly different from their actual values, life as we know it could not exist because complex structures would not exist. This is the first part of the anthropic issue: why are the laws of physics biofriendly (Ellis 2006; Ellis 2012)?

Second, there is the metaphysics of cosmology: the issue of what decides why the cosmos should be as it is (Level 9). Some boundary

Table 2

THE BIFURCATING SCIENCES HIERARCHY: FOUNDATIONAL LEVELS

Level 9	Metaphysics of cosmology	Morality, Telos, Meaning
Level 8	Cosmology	Sociology/Economics/Politics
Level 7	Astrophysics	Psychology
Level 6	Planetary/Earth science	Physiology
Level 5	Geology	Cell biology
Level 4	Material science/Mineralogy	Biochemistry
Level 3	Chemistry	
Level 2	Atomic physics	
Level 1	Particle physics	
Level 0	Laws of physics	

conditions are required for the universe: these also must be finely tuned, else there will be no stars and planets providing a haven for life. This is the second part of the anthropic issue: Why are the boundary conditions for the universe biofriendly (Ellis 2006; Ellis 2012)?

Now some recent books have tried to claim that the nature of the laws of physics inevitably leads to creation of an ensemble of universes out of nothing, and hence this is why a biofriendly universe exists (Hawking and Mlodinow 2010; Krauss 2012). This is then claimed to solve the major philosophical problems of existence. There are various problems with this: first, the supposed generation mechanism for universes is speculative in the extreme, and is certainly untestable as a scientific theory. Second, this claim is extremely naïve in philosophical terms. Something with a bit more depth is required.

The link to acknowledge that there is other evidence on the nature of the universe apart from that given by hard science: our individual and social lives provide evidence as well (Ellis 1993; Murphy and Ellis

1996; Ellis 2011). In particular, ethical experience is a crucial part of social and political life. This forms part of the higher level on the human side: morality, telos, and meaning are the set of concerns that shape how we choose all lower-level goals and hence guide what happens on the human sciences side in a top-down way. In particular they constrain lower-level goals according to what is acceptable or desirable and what is not, thereby having a key influence on what happens in the real world. Thus these are at the corresponding top level of the hierarchy of causation on the human sciences side.

Now a key issue is how to regard these latter aspects: Are they human inventions? Are they the inevitable outcome of the mindless workings of the Darwinian process of evolution? Or do they represent something deeper: a fundamental aspect of the way things are, in analogy with mathematics?

I will not go into the argument here, but rather simply state that, like Raimond Gaita (2004), Nancey Murphy and I argue for moral realism (1996; Ellis 2008). Curiously we are supported in this by some of the "new atheists," even though they do not seem to understand that they are doing so. The writings of both Richard Dawkins (2006) and Viktor Stenger (2007) make strong claims about the evil caused by religion. In doing so, they are presuming to make a claim that is more than just their personal opinion: they are expressing this as irrefutable fact. This argument shows a belief in absolute standards of right and wrong, independent of culture and space and time, which is in agreement with my own position that underlying the universe is a moral reality.

And of course that was Sir John's position, too. More than that, he believed the underlying principle of existence was *Agape Love* (1999), which he showed was at the spiritual core of all the major world religions. And this accords with my view and that of Nancey Murphy that *kenosis*—the spirit of loving self-sacrifice—is the universal core feature of ethics at all times and places in the universe, realized as such by the great spiritual leaders of all religions and cultures.

If that is the case, one can unify the three fundamental principles—the metaphysics of cosmology, the nature of the laws of physics (together allowing intelligent life to develop in the universe), and the meaning incorporated in morality and ethics and profoundly expressed in beauty—as all being expressions of underlying meaning and purpose—that is, in ordinary terms, they express the nature of God who underlies it all (cosmology, physics, and ethics) and gives it all its purpose and meaning.

This is nothing other than the message of St. John's Gospel as explained so eloquently by William Temple (1985), which fits together perfectly with current scientific interpretations of the anthropic principle in cosmology (Ellis 1993; Ellis 2008). One arrives at the view put forward in Table 3: The underlying principle that unifies it all by being the profound basis at both the top and the bottom is the loving, sacrificial nature of God.

Table 3

THE UNIFIED HIERARCHY: FOUNDATIONAL LEVELS UNITED

	Ultimate Reality: "God"	
Level 9	Metaphysics of cosmology	Morality, Telos, Aesthetics
Level 8	Cosmology	Sociology/Economics/Politics
Level 7	Astrophysics	Psychology
Level 6	Planetary/Earth science	Physiology
Level 5	Geology	Cell biology
Level 4	Material science/Mineralogy	Biochemistry
Level 3	Chemistry	
Level 2	Atomic physics	
Level 1	Particle physics	
Level 0	Laws of physics	
	Ultimate reality: "God"	

This provides a much more unified and profound view of the deep nature of reality than the strident proposals of the new atheists, whose reductionist standpoint denies all the deepest strivings and experiences of humanity through the ages (Atkins 1995). It is a view arrived at not by science alone but by a combination of science and humanity—philosophy at its best. I believe it is in full accord with Sir John's view of the world and the universe.

ROBERT JOHN RUSSELL, PROFESSOR OF PHYSICS,
UNIVERSITY OF CALIFORNIA AT BERKELEY

Unlimited Love as Ultimate Reality: Reflections
of a Christian Physicist on Sir John's Metaphors for God

Two of Sir John's favorite expressions for God, perhaps his most favorite ones, are "Unlimited Love" and "Ultimate Reality." He might even be heard to combine them in the phrase "Unlimited Love as Ultimate Reality." Sir John saw nature and the divine more in terms of continuity and discontinuity: the world, while remaining the world, is taken up into the divine being; the boundaries between matter, mind, and spirit are porous; and the invisible, both mind and spirit, best symbolize what is the incomprehensible Ultimate Reality. Sir John's views thus allow for great fluidity in the ways metaphors drawn from science can be used theologically since, ultimately for Sir John, they both study the same reality. Here in this short essay I want to follow Sir John's theology and try out a new series of metaphors for Unlimited Love as Ultimate Reality, metaphors drawn from key concepts in quantum physics and its remarkable account of the subatomic realm of nature.

The first concept is "superposition," a very surprising feature of the subatomic realm. In our ordinary world, most physical choices are discrete. You want to leave your room and there are two doors. You can go through one or the other—but not both at the same time. In the subatomic world of quantum mechanics, you actually go through both doors at once. So the states of going through one and going through the other "superpose" into the combined state of going through both (see the famous "double slit" experiment). This fact helps explain many of the basic properties of the chemical elements as represented by the periodic table. Quantum superposition underlies the way electrons in atomic orbits bind atoms together into molecules like water (H_2O) and salt ($NaCl$), it gives rise to the distinct line spectrum of light associated with each element, and it is the basis of the colors of paints, tinted glass, and dyed fabrics. Even the joint reflection and transmission of light is based on superposition, a phenomenon seen routinely as you sit by a flowing river and see both its surface and the rocks underneath it.

Superposition, in turn, suggests that the real nature of the world is something like a compound state of many distinct, even contradictory, possibilities. It is as though nature were able to be many things at once, an experience for us that seems present (perhaps only) in our mental world. Consider our ordinary experience of time as flowing from the future through the momentary present into the ever receding past. While we cannot literally go back in time, in our mind we can recall the past and anticipate the future, and we can do so such that both are simultaneously held in our present awareness—a mental phenomenon that philosophers and theologians from Plotinus, Boethius, and Augustine to Bergson and Whitehead called temporal thickness or "duration." We can remember traveling last year to Europe and to India at the same time. We can weigh future alternative actions like eating chocolate or dieting before deciding which action to take. In this sense quantum superposition seems to point to something vaguely "mental" about the most elementary level of nature. Perhaps the classical idealists—from Plato to Bishop Berkeley—were more correct than their opponents, the materialists, from Democritus to Hobbs. Quantum mechanics certainly invites us to reopen the question of whether nature is more like mind and spirit than like inanimate matter.

A second concept in quantum mechanics is nonseparability, as highlighted by Bell's inequalities. Nonseparability reflects an even more surprising phenomenon than superposition because, while it is based on it, it extends superposition to several particles at once. Think about two elementary particles: electrons will do nicely, although it could be photons, too, and so on. Suppose two electrons are initially bound together in a single state. The state then decays, and the two electrons race off in opposite directions along the z-axis, headed for labs A and B, respectively. Suppose we measure their spin along a given axis (such as the x-axis) as they arrive at labs A and B. We repeat the experiment for a second pair of electrons, then a third, and so on for a huge number N of electron pairs. The results of the N spin measurements in each lab look entirely random. But if we later compare the measurements from labs A and B, we find that every time lab A got one result, say "spin up," lab B got the opposite result, "spin down." We conclude that the electron spins are exactly "correlated" (technically they are anticorrelated, one up, the other down). Now we repeat the experiment, but this time we measure the electrons' spins along different axes: the x-axis in lab A and the y-axis in lab B. The results are striking: the correlations we find for measurements along these different axes contradict what we would predict if we think of spin as a property carried by each electron, a property like mass or charge. What does this mean?

There are two leading interpretations of this result: One is that there must be signals interacting between the electrons and traveling faster than light. These signals force the electrons into their correlated state. But this claim seems to violate Einstein's special relativity, which precludes causal signals traveling faster than light. Another idea is that the two electrons always manifest a single underlying state even though they seem to act independently when we study them separately in the two labs. I prefer this second interpretation, for it suggests that there is an underlying unity to nature, a holism if you like, which the explicit empirical phenomena (the individual results from labs A or B) do not portray on their own. French physicist and philosopher of science Bernard d'Espagnat, the 2009 Templeton Prize laureate, likes to say that nature is "veiled." It is as though nature's underlying ontological unity is hidden behind its discrete outward appearances even while something about these appearances, such as quantum correlations, hints at this unity. Again one can think in terms of mental metaphors more easily than in material ones about this unity. Perhaps the "ultimate character" of nature is more one of unity than of diversity or one of diversity within unity rather than one of mere chaos. Perhaps nature is more one than many, more utterly simple than compound, more elemental than complex, more spiritual than material.

When we take these concepts and their associated metaphors from quantum mechanics into the realm of theological discourse—where, as Paul Tillich stressed, all theological concepts are symbols pointing to what is truly real while falling short of depicting that reality—we may find rich new resources for expounding what Sir John means by calling God "Ultimate Reality." Surely God is hidden in incomprehensible mystery beyond the limits of human experience, whether spiritual, affective, or rational. Yet the God Christians attest to is a God who loves in freedom, who creates the world out of love for the world, and who makes Godself known in and to and for that world, especially in the lives and faith journeys of people as recorded in sacred scripture. Perhaps the superposition in nature we find through quantum mechanics suggests that mental and spiritual language about Ultimate Reality—God—is, after all, more appropriate as a fund of symbols for speaking about God than ordinary language drawn from the world of concrete objects and events. And perhaps the quantum correlations in nature again found through quantum mechanics point at least partially and imperfectly to categories of unity, the one, and the simple in nature. In doing so they can suggest a trace of nature as being created by Ultimate Reality, a trace not surprising for a nature created by One whom we believe to be true unity, truly one and simple.

Of course, all human language, whether based on personal experience or history—or quantum mechanics—is bound to fail to depict the true ultimate mystery that is God. Nevertheless I think these suggestions for metaphors drawn from mind and spirit over matter and for unity across space and time over separateness in space and time point in an evocative way to Sir John's claim that God as Ultimate Reality is love in its unlimited, unconditional, and endless expression, and that the relation between nature and God is more fluid and open than we traditionally have believed. Such a view of God affirms God as the ongoing creator of the world, a view that both respects God's ontological mystery while understanding the world as a veiled unity among explicit diversity, a world better depicted through metaphors of mind and spirit than of simple materiality. Once again, Sir John's vision about the deep relation between nature and God has provided us with an invitation for boundless further exploration.

IAN G. BARBOUR, PROFESSOR EMERITUS OF PHYSICS, SCIENCE, AND RELIGION, CARLETON COLLEGE

Reflections on Sir John and Unlimited Love

I came to know Sir John first from meetings of the scientific advisory board of the John Templeton Foundation (JTF) and then, after I had given the Gifford Lectures and received the Templeton Prize, through attending meetings of the members of JTF. I was impressed by the way he combined a deep commitment to the Christian tradition with an openness to other traditions. He always tried to have representatives of all the major world religions on the panel of judges for the prize. He promoted new JTF initiatives in Russia, China, and Eastern Europe. A succession of research projects on altruism and unconditional/unlimited love were among his top priorities.

Sir John grew up in a Presbyterian church in Winchester, Tennessee. That tradition has stressed the sovereignty of God, human sin, and divine justice and grace. It has believed in predestination—and usually in double predestination (each of us destined for heaven or hell). But he was also influenced by his mother's loyalty to the Unity Church, which puts a greater emphasis on God's love and human freedom, offering a more optimistic view of human nature.

I am the son of a Presbyterian geologist from Scotland and an Episcopalian mother from Brooklyn, but I have found myself more at home first among Quakers and then in the United Church of Christ (Congregational). I have been trying to rethink the relation between God's power, God's love, and the reality of evil. Why would a powerful and loving God allow the extremes of evil and suffering that we have seen in recent decades—six million Jews exterminated in Nazi concentration camps, or the violence and tragedy of conflicts in Iraq and the Mideast? God may value human freedom and respect the laws of nature in the created order, but was it worth the suffering to which they lead?

If God's power is emphasized, God's love seems to be compromised.

My younger brother died of a brain tumor at the age of twenty-six. After the memorial service several people said, "It must have been a comfort to know that it was God's will." They were trying to be helpful, but I thought: *No, I don't believe God planted that tumor.* Surely God grieves the loss and stands with me in my sorrow. One meaning of the cross is that in Christ, God participates in human suffering but does not let it have the last word.

So I am willing to consider some limitations in God's power in order to allow a larger role for God's love. In classical theology, God was said to be omnipotent, omniscient, and unchanging. But if human freedom is real, and chance and contingency are part of evolutionary history, even God cannot know the future. We cannot say that all time is spread out simultaneously before God, if the future does not yet exist. If God responds to events in the world, God cannot be unchanging. This would be a voluntary self-limitation on God's part because love is God's essential nature; it is not a limitation imposed by some external power. I have been indebted to the process philosophy of Alfred North Whitehead and to process theologians such as John Cobb. They portray a God who inspires and persuades rather than coercing us. Such a God is neither omnipotent nor impotent, but empowers others from within.

There are many differing images of God in the Bible and in subsequent theology. But as feminist authors have pointed out, the dominant image is of God as King and Sovereign—which is not surprising when almost all theology was until recently written by men. Moreover, the tradition has emphasized God's transcendence, though it has not neglected God's immanence. The Holy Spirit represented God's presence in the world. In the first verse of Genesis, the Spirit is moving over the face of the primeval waters. The Spirit inspires the prophets and supports the worshipping community, is present at the baptism of Christ and to his followers at Pentecost. I believe we should put more emphasis on immanence today, to balance the emphasis on transcendence in most of Christian history. God does not need to intervene supernaturally in violation of the laws of nature if the Holy Spirit is already present throughout creation. After describing the wonders of plant and animal life, Psalm 104 asserts (in the present tense), "When thou sendest forth thy Spirit, they are created."

Is altruism present in the world of nature? Evolutionary biologists have asked how self-sacrificial behavior could persist among social insects if their individual reproductive futures are jeopardized thereby. Research has shown that such behavior enhances the survival of close relatives who share and pass on many of their genes (in an ant colony, for example). Moreover, cooperation and symbiosis have been as important as competition and individual strength in evolutionary history.

Studies of chimpanzees have shown that they have strong social bonds and even a sense of justice (in the distribution of food, for example). They can learn rudimentary forms of symbolic communication (with hand signs or keyboards). But humans alone are capable of abstract language and the accumulation of information that culture makes possible. Humans alone can imagine alternative futures that can be evaluated in terms of long-term goals. These are all prerequisites of genuine altruism in which a person seeks the welfare of another even at some risk to himself or herself. Sometimes people do have ulterior motives and act generously to enhance their reputations, or in the hope that their generosity will be reciprocated. But such dismissal of altruism does not fit cases such as the families in Nazi-occupied France who risked death by hiding Jewish children and taking them over the border to Switzerland.

A human being is organized in a hierarchy of levels studied respectively by physics, chemistry, biology, neurology, and psychology. *Reductionism* is the thesis that what goes on at any level can be explained by the laws governing lower levels. According to reductionists, causality operates from the bottom upward. We are determined by our genes and by the neurons in our brains. By contrast, *emergence* is the thesis that distinctive concepts and theories are required to explain phenomena at higher levels. Events at higher levels can influence events at lower levels, not by violating lower-level laws, but by setting boundary conditions for their applicability.

Systems thinking has been fruitful in many areas of science today.

Complexity theory shows that feedback loops between levels can generate new phenomena not anticipated by the theories of either level. In ecological systems the interaction between members of two species are tied into a whole network of other species. These views are *holistic* in directing attention to whole systems, without denying the importance of studying the separate parts and how they are organized.

In place of the body/soul dualism dominant in Western theology, and the matter/mind dualism prominent in philosophy since Descartes, many theologians and scientists are viewing a person as a psychosomatic unity, an embodied and socially extended self. Mental phenomena such as feelings and intentions involve whole networks in the brain and the rest of the body. In such a framework it is possible to defend free will and responsible choice for which we seek reasons rather than causes. These are prerequisites of moral responsibility and genuine altruism. We have few clues as to how consciousness and subjectivity arise, or how first-person accounts of mental events relate to third-person accounts of brain activity. Perhaps we should accept e*xplanatory pluralism*, the thesis that explanation has differing goals in differing contexts.

I would conclude that the biological roots of love and altruism lie in prehuman life, but unconditional love seems to be possible only for God and as an ideal for human life. Sir John's vision of unlimited love can continue to inspire theological reflection, scientific research, and the interaction between these forms of inquiry.

REBEKAH DUNLAP, UNITY SCHOOL OF
CHRISTIANITY MINSTER

Unlimited Love as Ultimate Reality

Sir John Templeton was one of the most remarkable persons I had
the honor and privilege to know, and for our writing partnership and
friendship, I shall be eternally grateful. Sir John was known worldwide
as a pioneer in financial investment and as a dedicated philanthropist
for humanity. The extraordinary breadth of his career included a magnificent vision that united science, religion, and philosophy. After Sir
John officially retired from the money management world of finance, he
directed his main focus toward advancing spirituality and improving the
human condition. The John Templeton Foundation, based in Conshohocken, Pennsylvania, under the leadership of his son, Dr. John "Jack"
Templeton, provided the vehicle to achieve this goal.

Sir John and I initially met at an Association of Unity Churches conference in Lee's Summit, Missouri, while I was serving on the executive
board of said organization. I knew of Sir John through his work with
the Templeton Funds for global investing, which was later purchased by
Franklin Investments and became the Franklin Templeton Funds. Sir
John's provision of research grants in the fields of medicine, science, and
religion was also well-known. I began working as a writing collaborator
with Sir John in October 1994 when the invitation came to revise and
update his book *Worldwide Laws of Life*. He had read a couple of my
previous books and felt the writing style was compatible with what he
wanted for *Worldwide Laws of Life*. Thus began a special friendship and
deeply meaningful collaborations in preparation for the four additional
books that unfolded over the next ten years. When the writing assign-
ments were completed, we remained in touch as friends.

My first meeting with Sir John for writing purposes occurred at his
home in Lyford Cay, a small community of Americans and Europeans
in Nassau in the Bahamas. My plane was arriving near midnight, and
since Sir John would be in a late meeting with a contingent from Japan,

he recommended I take a taxi from the airport to his home. He told me where the key to the guest house would be in case I arrived before he returned. As the taxi pulled up the winding drive to the front of White Columns, two large white German shepherds came loping across the lawn and stopped beside the taxi. The driver looked at me and said, "Lady, I'm not getting out of this cab."

"Not a problem," I responded. I handed the driver a tip, opened the door of the taxi, stepped out onto the drive, removed my one piece of luggage, and stood quietly looking at the dogs. I held out my hands, palms up, as they did a walk-around inspection, and then I walked to the front door with an escort on either side! As I started to ring the doorbell, Sir John opened the door, observed the scenario before him, smiled, and said, "Oh, my!" then reached out his hand in welcome. He guided me to the guest house next door where we sat in the living room and talked for a while. As he left that evening, Sir John advised that fresh oranges and grapefruit were available on the trees outside the kitchen door should I desire them for breakfast!

Our initial conversation confirmed what I had heard and felt. Sir John was a visionary in many ways who held a strong belief that religious values are basic and intrinsic for human liberty as well as for the free enterprise system. His wisdom provided tremendous insights into the human condition worldwide, and his spiritual dedication and commitment were directed toward helping humanity. His writings, financial contributions, and the ongoing work of the John Templeton Foundation provide continuous exploration in the fields of spiritual, scientific, and philosophical endeavors.

Working with Sir John encompassed an exercise in expanding spiritual awareness since the majority of our conversations flowed along this focus. Especially intriguing is his motto for the foundation: "How little we know, how eager to learn." Sir John believed that "there is a definite difference between *acquiring* knowledge and information and *possessing* wisdom. A person may acquire knowledge and meaningful information from attending a university, through travels, through relationships, through books that are read and studied, and through a variety of activities in which one may participate. But is the person also gaining wisdom?" (1997, xxii). How can one accept on hearsay that which he knows nothing about? He emphasized the importance of *living* what you believe until it becomes an inner knowing. Throughout his life, Sir John encouraged open-mindedness and wrote extensively on spirituality and the role that scientific research could play in expanding the spiritual horizons of humankind. Exploring new avenues prepared a strong and workable foundation for attaining his many goals.

Numerous faxes over the years and occasional trips to Nassau offered opportunities to discuss a variety of topics in the fields of science, religion, and philosophy. Sir John felt strongly that there must be a unifying thread of truth that linked these vast fields of knowledge and exploration in a unity of purpose and expansion. Having served as a Unity minister for almost two decades when I met Sir John, I was especially interested in, and intrigued by, our conversations along spiritual/religious perspectives. He felt that a powerful, enlightening, and helpful body of sacred wisdom is available through the writings and tenets of the great religions of the world, and an individual could benefit expansively by exploring these avenues. The rich variety of the major world religions creates a tapestry of amazing beauty—a testimony to the essential spiritual nature of our human visit on earth. He wrote in the Introduction of *Wisdom from World Religions: Pathways to Heaven on Earth*, "Evidences are increasing that there's more to why you're here than what you presently know! Your search for that 'something more' can be a sacred adventure into new insights, provocative choices, unexpected turning points, and an enthusiasm to enter the unknown." I asked Sir John once what he meant by "sacred adventure." He responded that we are essentially spiritual beings. Our world is basically a spiritual world, and the underlying operational forces that keep everything working in harmony are identified as spiritual Laws or Universal Principles. When we establish conscious unity with this spiritual essence, we begin to recognize that the visible and tangible are only minute manifestations of the vast timeless and limitless reality. All aspects of creation are in an evolutionary process. Our earth is a school, providing unlimited opportunities for experience and growth. Our evolutionary journey, as we grow in spiritual understanding, becomes a sacred adventure in which one can express actively and fully in daily life the inherent spiritual qualities. Recognizing the sacred aspects in, and of, everything we do brings additional benefits to every aspect of daily living. Sir John often noted the vital importance and significance of taking time to be in quiet communion with God—by whatever name the Absolute, Ultimate Reality, the Creative Source, God, Allah, the Christ, the Buddha, and so on may be called. Each of us is a spark of that One Life that irradiates and sustains all manifestations, and therefore only true happiness consists in the thoughts, feelings, and actions that are consistent with the unity of all life.

In our collaborations, Sir John used the terms "God," "Creator," and "Ultimate/Infinite Reality" frequently interchangeably. However, we both knew the subject is an "Omnipresent, Eternal, Boundless, and Immutable Principle" for which humanity has no name and upon which all speculation is impossible since it transcends the power of human conceptions. "God" includes, of necessity, the temporal and partial as well as the eternal and boundless. We examined the consideration that wherever we looked, science, religion, industry, education, psychology, and other fields of endeavor were exploring broader questions such as: Are we poised on the verge of experiencing a greater Reality? What is this greater Reality? Does Ultimate Reality indicate a higher order in the universe that cannot be explained by presently known physical laws?

Provocative questions often energized our discussions as our imaginations took us into the realms and regions of Unlimited Love and Ultimate Reality—two topics of major importance to Sir John. After these stimulating occurrences, my mind would be spinning with the depth and degree of our discussion! Sir John emphasized the importance of a person's thinking process and encouraged "digging a little deeper," to expand beyond one's present level of thought, and to discover and recognize one's personal convictions regarding Universal Truth. He described this Truth in philosophical terms as the Ultimate Reality; in religious terms as the Mind of God; and for each individual as the supreme Mystery of Life. If we are to discover "Truth" itself, we must build into such a framework the flesh (substance) of living experience. Truth can only be *realized* as it is studied, applied, and expressed in daily life. Sir John felt that questions can be an invitation to greater awareness as they often point toward areas of our experience that need attention. When we allow the question that is implicit in our difficulty to become explicit, we are inviting our inner awareness to explore the situation and offer guidance.

Are we willing to make a commitment to the search for that ultimate reality that is mysteriously also our own deepest self? What is the importance of recognizing the presence of the sacred within us and around us? And what is the "presence of the sacred"? Could a starting point be an avenue of spiritual research that could provide a realization that the universe and our purpose in it could be much more than a haphazard occurrence? Nature is governed by law, and reason and logic tell us that this law must apply everywhere in the universe. Could we perhaps begin this research with a review and understanding of the Law of Reciprocity—or, as it is sometimes called, "cause and effect"? Along every step of the evolving spiral of consciousness, does the thread of truth become stronger and more radiant with every progression in life? As we evolve, the refreshing morning sunlight changes into the brilliance of noonday glory.

Sir John talked and wrote extensively on Unlimited Love. Why? Because he felt strongly that unlimited love can eventually assist humanity in arriving, through individual efforts, to that state of spiritual awareness and unity with all life that may lift us from the human to the divine—Ultimate Reality! How did Sir John define Unlimited Love? In *Why Are We Created?* we wrote, "By definition, 'unlimited' means without limit of restriction. Unlimited love can describe love for humanity and perhaps, on a less ontological level, love for all living creatures. Unlimited love can mean we allow no insulating boundaries to be drawn to separate us from others" (2003, 88). Unlimited love is a title free of association with any one religious faith tradition and can appeal across cultures and academic disciplines. It can be representative of the ultimate nature of love as a spiritual and creative influence. "Unlimited love is a principle quality that opens our hearts and minds to a new beauty. Moreover, open hearts and minds can be passages into evolving consciousness, individual and collective. Spiritual progress reflects growth in the conception and expression of spiritual ideas" (2002, 300).

Sir John often referenced the Greek definitions of love. *Eros* is romantic love, the kind that puts butterflies in your stomach. *Storge* is the type of love we feel for members of our family and is the love of security. *Phileo*, or companionship, is the type of love we feel for our friends. The most important aspect of love, however, he felt, is *Agape*, which is the unselfish love that grows as you give freely and expects nothing in return. It is said that no power in the universe is greater than love, and no act more important than loving. Unlimited love deepens our understanding and enables us to discern from a wider perspective. Beauty may be discovered in unexpected places as the light within recognizes and discerns the light of truth eternally shining through the broken shapes of the temporal.

In a letter dated April 11, 1998, Sir John wrote,

> Help me . . . prepare an article of about 5,000 words called, "Unlimited Love." I would like the article to be equally attractive to Muslims and Hindus, as well as Christians. I would like for most of it to be written in the form of questions and possibilities rather than certainties or dogma.
>
> Maybe there is a happy middle ground for most endeavors, but there can never be any excess in learning to love every human being without any exception and to feel and radiate not a little love, but unlimited love. This does not need you to admire or initiate or give material help, but only to *feel* and *radiate* unlimited love. Love may be the basic spiritual motivator which can lead to forgiveness and giving and thanksgiving and prayer and other spiritual qualities. Maximum love for your child does not mean that you relieve him from work and tests and self-reliance.
>
> The more love you give, the more you have left. The more

you radiate love, the more love will return to you. Love eliminates conflict, anger, resentment, and most other mental and spiritual poisons. Love may be a universal force more potent than gravity, electromagnetism, or light. Love may explain why humans were created and how we are expected to be agents in accelerating creativity.

Then came the questions:

Why and how is unlimited love an essential ingredient in living a successful, beneficial, and happy life? How can the extraordinary potential of unlimited love be recognized and realized? Is not love an inner quality that sees good everywhere and in everything and does not get caught up in faults—imagined or otherwise? Our stay on this small planet called Earth is a brief one and we have an excellent opportunity to leave the world a better place than we found it through our choice of how we live our lives. Is one way to accomplish lasting improvement that of mastering the Laws of Life, and the greatest of these Laws is Love? Is unlimited love possibly a primary purpose in life? Could our souls have been brought forth in love and could this love foreshadow our purpose on earth? Are the love, kindness, and patience that we bring to our relationships ways to be in touch with spiritual discoveries? Can unlimited love be described as a creative, sustaining energy? Does a divine fountainhead of love exist in the universe in which degrees of human participation are possible?

"One fundamental of psychology is that a great need of humanity is to be loved. We may nod our heads in agreement with that thought. 'Yes, I agree. Life can seem empty and meaningless unless we're loved.' Do we need to look at this concept from a different perspective? Could it be that the most important way a person can be loved is to *give* love? As long as we're looking for love outside ourselves, we may be frustrating the love inside!" (2002, 13). Unlimited love is a natural attitude and demeanor of good will, kindness, compassion, caring, support, and benevolence. It is a willingness to do what you can to be helpful and to make things a little better for someone. When we allow the eyes of the heart to be opened in unlimited love, we see more of the realities hidden behind the outer forms of this world. When the ears of the heart are open, we hear more words of truth that may be hidden behind other words.

Sir John endorsed a humble approach to life and asked, "Do we radiate love and happiness as faithfully as the sun radiates light and warmth?" Unlimited love and humility—are they connected or related? Releasing egotistical self-will and inviting the richness of divine and unlimited love into our lives is important.

A discussion in October 1999 focused around the topic of Ultimate Reality. Again, our process utilized questions as food for thought as we delved into this topic. Is there evidence that human perceptions of reality are accelerating? Could we discover more if we use the word "reality" to mean the totality of appearances plus fundamentals? We may investigate what reality is like, what it is not, and speculate on how we may discover more. Is our language inadequate to describe the scope of reality. Does "reality" have dimensions? Does a vast realm of intelligence exist beyond human thought, with thought representing only a tiny aspect of that intelligence? How do we account for the "layers" of infinite intelligence—not yet discovered by humans—beneath the surface of everyday phenomena? Are we now recognizing there may be multiplying evidences of purpose in the universe and in creativity?

At this time, Sir John was working on a book draft titled *Possibilities* and sent a fax on October 29, 1999. "I am attaching to this fax my draft of a chapter (Chapter XIII—"Infinite Intellect") for my new book *Possibilities*. I would be grateful if you can fax back to me, within a few days, suggestions for improvements or additions to this text."

Powerful questions arising from this discussion were as follows:

Do Ultimate Reality and Infinite Intellect go hand in hand as two aspects of the same whole? Are there other infinite intellectual worlds both like and unlike this world of ours? Is this simply speculation, or how could we *know*? How does Infinite Intellect assist each person in playing his or her personal note in the universal symphony of life? Do we carry within us an innate sense of the Infinite Intellect and become inspired to think, feel, and do the things that maintain harmony and productivity within us and within our environment? How can this be brought forth in conscious awareness? Does this have anything to do with "intuition"? As we "remember," recognize, or understand our oneness with the Infinite Intellect, can we move into a greater energy flow of creativity? What is Oneness? Is Infinite Intellect a "seed atom" in a person's consciousness? How does one "discover" a larger fraction of Infinite Intellect? How does Infinite Intellect enter into the statement, "To learn the truth about the great cosmic universe, first learn the truth about yourself"? Can we find gleanings of this truth in the ancient wisdom teachings of the world religions?

One day, a question and discussion session regarding the human ego and its pros and cons stimulated exciting ideas. Are we often held back by the limitations of our own mind and narrowness of our cultural experiences? Do we see dimly the vast unseen and understand less of the Creator who lies behind all that is? Are we deluded by our egotistical mortal processes into thinking we have a fairly defined understanding of life and "consciousness"? Could we write a special fable of discovery, a magical story of an egotistical clam who thinks he and his community are the center of the universe? Could our human perception of reality be as meager as a clam's perceptions of humans? Could we guide this clam fellow through a series of events to help him become aware that the world is greater than he had thought and that joy, love, and creativity exist within us all?

In a letter dated September 13, 1996, after discussing the idea for the book project that became *Story of a Clam*, Sir John wrote, "Perhaps we could present a fable of scientific and spiritual correctness, with a clam as the 'main character.' We could show how egotism caused him to consider himself the highest and final form of life. Our book could show that humans are egotistical if they think they know more than a tiny fraction of the reality of the magnificence of God. My thought is that you might help me clarify this concept." Thus began the birthing of Columbus Clam!

Unlimited Love, Ultimate Reality, and Infinite/Ultimate Intellect—are these three aspects of a reality far beyond our present comprehensions? Sir John spoke of unlimited love as ultimate reality. He felt that enough pure, unlimited love could open the door for greater understanding of all of life's principles. What could happen in our world if humanity began to recognize that we are parts of something greater than ourselves, and that much can be gained for everyone through cooperation and unlimited love? When we realize that what we do affects others and what they do affects us, the phrase, "We are our brothers' keeper and he is ours," becomes deeply meaningful. Sir John knew and practiced this great truth. He set an example of living his beliefs, and look what he accomplished!

MARY ANN MYERS, SPECIAL PROJECTS,
THE JOHN TEMPLETON FOUNDATION

Sir John: The Evidence for Love

Sir John's achievements as an investor and a philanthropist were grounded in his respect for evidence and his diligence in seeking it. But his focus on scientific discovery that might speak to the question of God's existence or God's nature did not reflect agnosticism on the subject of divinity. Sir John had in hand the critical evidence—and it was love. The love he received and the love he gave: not linked to merit but unconditional, unlimited, unceasing.

His published writings are a testimony to his conviction. But for the purpose of these comments in response to Stephen Post's appeal, I would like to focus on his first book, *The Humble Approach*, which I was given when I joined the John Templeton Foundation in 1997. Subtitled *Scientists Discover God* and written in 1981, it was subsequently expanded and a new edition published two years before I met Sir John. In the first edition's penultimate chapter, "Love and Happiness: The True Test," Sir John writes, "God loves us all equally and unceasingly. It is His nature to do so. We should seek always to let God's love shine forth like the light inside an electric bulb illuminating all our habitation." (1981/1995, 110)

He then goes on to quote the Swiss Protestant theologian Emil Brunner (1889-1966): "In Jesus Christ . . . the Creator reveals Himself as the One who has created us in love, by love, and for love" (1932/1948). The citation is from *The Divine Imperative*, which was translated into English in 1937, two years before Brunner and his young friend, the Scottish theologian Thomas Torrance (1913–2007), who would go on to win the Templeton Prize in 1978 and serve on the founding board of the foundation, made a joint decision to return to Europe from Princeton Theological Seminary (PTS) on the eve of the Second World War. Brunner's work, which was translated into English far earlier than that of his fellow Swiss theologian Karl Barth, was very influential in

American Protestant seminaries through the early 1950s, when Sir John first joined the PTS board. Brunner's influence faded as Barth's writings became available in translation later in the decade and came to dominate mid-twentieth-century theology. But both men wrote in reaction to a liberal Protestantism that reinterpreted fundamental Christian doctrines from an anthropological viewpoint that minimized the role of a transcendent, self-revealing God.

I do not know if Sir John was aware of Barth's famous "*Nein*" in response to Brunner's attempt "to find the way back to a true *theologia naturalis*" by insisting on the ability of human beings to discern God in nature (Brunner 1934). But Torrance may have discussed it with him, as Sir John engaged Torrance to edit the draft of his manuscript for *The Humble Approach*, whose main theme, he said as he sought aid in correcting errors and making his ideas "more cogent and readable," was that "by humility a new kind of theology can be built which can never become obsolete" (John Marks Templeton, letter to Thomas Torrance, July 15, 1979). In any case, I find it significant that Brunner is the only major theologian Sir John quoted in his little book. One finds scientists like Einstein and Carl Sagan and engineers like Vannevar Bush there, along with the transcendentalist preacher Theodore Parker, the Scottish evangelist Henry Drummond, the New Thought writer Ralph Waldo Trine, and the religious studies scholar Huston Smith. But the only other theological voice is that of Arend van Leeuwen, who is also a historian and is quoted on human passivity in a passage on creative genius.

Brunner wrote, "God . . . leaves the imprint of his nature upon what he does," and Sir John's agreement is suggested in his question: "If [it is true that 'God is Love'] was all of creation brought forth by divinity expressing itself as love?" (Brunner 1934/1946, 25; Templeton 2002, 199). I suspect the Brunner quotation was deliberately chosen to express a love-centered theology with which Sir John found himself very much in sympathy. And it also hints at Sir John's strong resonance with the idea that science, philosophy, and theology can meaningfully converse on big questions in search of truth. Sir John was open to surprises.

It may be that that openness, indeed, that *courting* of surprise played some part in his delight in talking to people from a vast range of business and professional backgrounds. But his *attentiveness* to them, as I observed it, reflected more than a search, ongoing as it was, for information more so than courtesy. It reflected his love for others. Sir John's kind of love was not sentimental, and he never engaged in flattery. Still, the bright, intent gaze, the sparkle in his eyes, the hand cupped behind his ear, the arm extended around a shoulder, were physical manifestations of unselfish love. He defined that kind of love simply as "a willingness to do what you can to be helpful" 2002, 13). It was an emotion he acknowledged needed to be cultivated, but the ability to love purely, without keeping score or counting the cost, he thought "becomes the purpose for our existence," never "fully achievable in this life"—nevertheless, "one of the greatest forces on earth" (Templeton 2002, 13; 2000a, 161; 2000b, 22). Sir John radiated love, and the love he imparted, more even than the love he encountered, gave him joy.

RUSSELL STANNARD, ADVISOR,
THE JOHN TEMPLETON FOUNDATION

A Higher Love and Evolution

Your pet cat goes out into the garden, sees a bird, pounces, and kills it. This despite you having recently given it a bowl of pet food so that it cannot possibly eat another mouthful. The reason for this apparently wanton, cruel behavior is that the cat cannot help itself. It is a type of behavior pattern encoded in its DNA. In the evolutionary "struggle for survival" it was an integral part of the survival pack of the cat's ancestors. Those animals that happened by chance to be endowed with an innate tendency to act in this manner were the ones more likely to catch prey. Rivals lacking such an instinct would have had to work out afresh each time that what they were presented with was a potential meal—by which time the bird had flown. So the former had a better chance of surviving to the point where they could mate and pass on their advantageous gene—the same gene we find in today's pet cat, even though it is no longer required (since the invention of kind owners). Much of the behavior of animals can be explained in this way. We say that the behavior of animals is *genetically determined*.

So does any of this apply to us humans? After all, we are ourselves a form of evolved animal. The answer most certainly is yes. Not that our behavior is genetically *determined*. Unlike the cat we do not have to helplessly follow our instincts. We are self-conscious. We can reflect on our actions and, if we so choose, decide to go against such innate tendencies. Nevertheless, it seems only reasonable to assume that our behavior is genetically *influenced*. What is written in our DNA inclines us to behave in certain ways—ways that were conducive to the survival of our ancestors. Clearly we have here an explanation of human selfishness, putting oneself first, looking after one's own interest—grabbing what's going. It can explain how some people show aggressive tendencies. It accounts for the desire to win—to excel over others in competition.

So, what about love? Can the phenomenon of love be explained (or explained away) in this manner?

Certainly it takes care of "making love." Those with a strong sexual desire for the opposite sex are clearly at an advantage when it comes to passing on one's genes to the next generation. It is when the libido is low, as it appears to be with panda bears, that the future of a species is put in jeopardy.

But what about other forms of love—those that do not have sexual connotations? What are we to make of those?

A mother pigeon is guarding her young. On seeing the approach of a hawk, she leaves the nest and makes a great display. This attracts the attention of the predator toward herself—and away from the helpless young. She is sacrificing her own life on behalf of her offspring. In humans we would regard such selfless behavior as an example of a mother's boundless love for her children. There is, however, in terms of evolutionary theory, a simple, down-to-earth explanation for such behavior. In the case of the pigeon, she has already passed on her genes to her young. What is important from the point of view of the propagation of the genes is that they survive in the young—the next generation. The mother has done her job. Provided that the young have reached the stage where they have a chance of surviving without her help, she is, in a sense, expendable. A gene that dictates that, when it comes to situations where it is a matter of you or your offspring surviving, you should choose the action that favors the offspring—such a gene has a better chance of surviving than one that leads invariably to putting one's own interests first. This explanation, involving the preservation of the so-called selfish gene rather than that of the individual, is known as *altruism on behalf of close kin*.

And what is true of pigeons is also likely to be true of us humans. It can be argued that evolution is able to account for mother love and any other acts of love we commit on behalf of those who are closely related to us and share to a large extent the same genetic material as ourselves.

But what of the love we have for fellow human beings in general? After all, we do not, as a general rule, go around robbing people, beating them up, killing them, raping them, and so on—behavior one might expect if genetically influenced behavior was solely about being aggressive, and selfish, and putting one's own interests first. Most people behave decently toward each other. We respect their rights. We extend brotherly love to those who cannot be regarded as close kin.

Again, evolution comes up with its own explanation. Take, for example, the behavior of monkeys. They groom each other. It is a case of one individual helping another, at some inconvenience to itself. Why behave in such an altruistic manner? Because the monkey needs to be groomed itself. It grooms the other on condition that the other will return the favor. Evolutionary biologists refer to this as *reciprocal altruism*. You scratch my back, and I'll scratch yours. It might equally be regarded as enlightened self-interest. Grooming is but one example of how situations can arise where there is advantage in cooperating with others, rather than competing against them. Hunting in packs, rather than trying to go it alone, would be a further example.

Thus, for us humans it is to everyone's mutual advantage to observe certain codes of conduct, such as: I won't take what belongs to you provided you respect my belongings; I won't attack and kill you, if you allow me to live in peace; I won't try to steal your wife/husband, if you'll leave mine alone. Here we have a ready explanation, in evolutionary terms, as to how one might account for three of the Ten Commandments. Call it "loving one's neighbor" if you like, but there is no getting away from the fact that it is to everyone's advantage if evolution were to endow us with a gene that inclines us to form societies of mutual cooperation and respect.

So we return to our opening question: Can evolutionary theory explain love? Yes and no. It can explain the forms of love we have so far dealt with. But there is another: unconditional love, or *unlimited love*.

As an example of such love, consider someone who makes a large donation to help the starving in Ethiopia. Why act in such a manner? Ethiopians are not close kin, so the selfish gene idea does not help. Moreover, the Ethiopians are never going to be in a position to return the favor, so it cannot be an example of reciprocal altruism. It can be argued that it makes the giver look good in the eyes of other people. It enhances one's status in society. Fair enough. The desire to be well thought of is undoubtedly a motivation in some cases. But what of the person who gives anonymously—as so many do? What is in it for them? Nothing.

One might counter by saying that it makes one feel good so to act, and that in itself is the reward. Yes, it can make one feel good. But is that an explanation? *Why* should one feel good? Why isn't it more natural to feel stupid? If one is lucky enough to be born into a prosperous country and enjoy a high living standard, why not make the most of one's good fortune rather than fritter it away?

Some might argue that one behaves in that way because one would hope others would act toward oneself like that if the roles were reversed. In other words, do unto others as you would have them do unto you. But where did that idea come from? Why imagine putting yourself in their position when you are not in their position, and are unlikely ever to be in that position? Again, why not cash in on your good fortune and advantages in the same way that your evolutionary ancestors made the most of whatever advantages they happened to possess?

This unlimited or unconditional love was the kind of love Jesus advocated, and he took it to new heights. "Love your enemies," he declared. Just imagine one of our ancestors getting a gene mutation that amounted to the instruction to love one's enemies. How long do you think he would last before he became a free lunch for some lucky predator? Then one has Jesus saying how there is no greater love than to lay down one's life for another. In this regard note he was not talking about laying it down just for close kin. He was himself to lay down his life for us—we who cannot be counted as his close kin.

There is no denying the existence of unlimited love. Each day we have countless examples of it. But how are we to account for it? Unlike those other types of love, it seems difficult, if not impossible, to explain it in terms of some by-product of our evolutionary history.

Christians hold that it comes from God. It is a gift from God. Only so could we be raised above the level of a mere evolved animal. Jesus taught, and himself demonstrated throughout his life, and especially on the cross, that it is a reality that transcends all others. Unlimited love is not just a feature of life—one that takes its place alongside the others that make up human life. It appears to be what this life is ultimately all about.

This is the fundamental truth that shaped the mind of Sir John Templeton as I knew him over many years.

HOLMES ROLSTON III, TEMPLETON PRIZE WINNER,
ADVISOR TO THE JOHN TEMPLETON FOUNDATION

Unlimited Love and its Limits

Agape love is central to Christian conviction, embodied in the life,
death, and resurrection of Jesus, and the iconic virtue in 1 Corinthians 13. Sir John Templeton, perhaps the most visible wealthy Christian
philanthropist in his lifetime, sought to realize such agape. For that he
is justly to be praised, a role model for us all.

So one hesitates even to ask if there are limits to unlimited love.
Still, we do need to think about the logic of love, both generally and
in the wisdom of John Templeton. Love is the cardinal virtue, but love
is not the only virtue, or duty. Neither in deontological ethics nor in
utilitarianism, the two main Western traditions, is altruism the pivotal
principle. The moral agent does what is just, giving to each his or her
due, and whether this due is to self or other is secondary. The question
of fairness (justice) is not so much one of preferring self over other (I
win; you lose), or other over self (you win; I lose), but of distributing
benefits and losses equitably (summing wins and losses, we each get
what we deserve). The agent does the greatest good for the greatest
number, which might mean benefits to self and/or to other, depending
upon options available.

The Golden Rule urges one to love neighbor as one does oneself, but
this is not other love instead of self-love. "Do to others as you would
have them do to you" seeks parallels in the self doing for others with
others doing for the self, suggesting reciprocity as much as antithesis between self and other. The first and most widespread Hindu and
Buddhist commandment is noninjury, *ahimsa*, whether the injury is
to others or to self. The commandment enjoins self-defense as well as
defense of others threatened with injury. Aristotle recommended the
golden mean, also a balancing of values. Doing the right, the good, is a
matter of optimizing values, which often indeed means sharing them,
but this is never simply a question of always benefitting others instead

of oneself. Socrates's concern is amply for the self doing well as the self does well by others. There is no egoism-altruism dichotomy pivotal to his ethics.

In John Templeton's writings, he admires wisdom of diverse kinds from multiple religious (and secular) sources, particularly those that support his humility approach to life. He loves aphorisms that support his laws of life. We hardly know more than an ant crawling along a shelf in the Library of Congress, he once said. So he devoted his wealth to enlarging human knowledge.

The Templeton Foundation has supported many initiatives with beneficial effect: the science/religion dialogue, meaning in evolutionary natural history, the evolution of ethics, the anthropic principle in cosmology, character formation, studies in the laws of life. One concern has been whether charity undermines the capacity of the beneficiaries for self-support—and here Templeton argues that free enterprise should be promoted in developing nations. All of this is commendable philanthropy, but it would be a stretch to think of most of this as agape, or unlimited love. Rather, Templeton invests in research, scholarship, and education important to its agenda.

The Hebrews claimed that the righteous person is "like a tree, planted by streams of water, that yields its fruit in its season," by which the sages, prophets, and rabbis meant both good deeds and a prosperous family. Such a person is, in their idiom, "blessed" (benefitted), and by contrast sinners "perish" (Psalm 1). The Hindus and Buddhists interpreted the value of virtue in terms of good karma, deeds that benefit others and self at once. Calculating whether the self wins or loses in a direct tradeoff with whether others gain or lose can hardly be said to be the principal axis of analysis of any ethical system in the classical past or contemporary present. The questions are more those of justice and love, or integrity and virtue, or honor, or of optimal quality of life—that is, of good and evil, right and wrong.

Many dimensions of morality do not directly focus on altruism: questions of the rights of the minority, of capital punishment, the extent of free speech versus pornography, preferential hiring, abortion, euthanasia, fair wages, and so on. Ethics is about optimizing and distributing moral and other values, about what sorts of values count morally, and what the moral agent ought to do to promote these values. This is a more comprehensive question than whether the self is preferred over others or vice versa.

Well, Christians may reply, just this shows the deepening of Christian conviction: agape, over *philia*, over *eros*. Jesus embodies suffering, sacrificial love, which is a level of concern unreached by Socrates or Plato, or the utilitarians, or advocates of human rights. Here the good is less than the best. God is love. God saves by grace alone, through faith. The issue is not merit, rights, justice, fairness. The thrust is forgiving love. Redemptive love in Christian discipleship exceeds more calculating loves; agape is unlimited love. This is the ultimate role model. Didn't Augustine say, "Love, and do what you will"?

Classical ethics, perhaps strengthened by classical Christianity, invites altruism and constrains egoism. Altruism in the ethical sense applies where a moral agent consciously and optionally benefits a morally considerable other, without necessary reciprocation, motivated by a sense of love, justice, or other appropriate respect of value. But in turn that requires that religion be concerned with more than altruism. Religions also are concerned with justice, fairness, equitable sharing of resources, prudent care of oneself, a right relationship to God, reaching nirvana or union with Brahman, and so on. In the Judeo-Christian tradition, the goal of forgiving grace is often said to be a state of righteousness. Altruism needs complementing with justice.

At this point, an inclusive, comprehensive ethics may choose to argue that what the impoverished, the poor, the downtrodden are entitled to is not so much charity as recognition of their human rights. They do not so much wish to be the ongoing beneficiaries of super-altruism as to receive fair treatment from those who have exploited them and who have perhaps become wealthy as a result of their exploitation. Doing the right is a matter of recognizing entitlement as much as giving gifts. Charity is voluntary, but such entitlements can and ought to be enforced in the courts, written into legislation, regulated, policed. Waiting for the philanthropic wealthy to fix the ever-increasing inequities between the rich and the poor looks in the wrong direction for a solution to the most pressing moral issue on the world agenda.

John Templeton delighted in being a contrarian. That was a role he recommended. But when I tried that within the Templeton environment, I found that John Templeton and his associates, while they seemed to listen, never took my concerns with ongoing seriousness. Of course, Templeton and those who worked with him to distribute his money wisely were deeply concerned that such philanthropy be effective, that it raise the standards of living of the beneficiaries, that they become more virtuous, fair, thrifty, self-sufficient. Amen, again.

But the Templeton agenda could never seem to register the need for structural reforms to the inequities of global capitalism. That will have to be addressed by other ethicists who face these problems and, also in love, confront global free enterprise with limits. Meanwhile, John Templeton exemplified unlimited love, superbly.

My impressions are based on a half dozen encounters with John Templeton, first at a conference on Empathy, Altruism, and Agape, at the conclusion of which he sought advice from the Templeton Foundation International Advisory Board; October 1999 in Boston; equally on discussions at that board's annual meetings in Nassau, from 2006 onward. He has, of course, written frequently on these issues. He heard me lecture on altruism once, in Philadelphia, April 1999, with some conversation afterward. Also, he spoke at the press conference in New York, 2003, when my winning the Templeton Prize was announced.

GLENN R. MOSELY, PhD; PRESIDENT-CEO EMERITUS, ASSOCIATION OF UNITY CHURCHES INTERNATIONAL

Unlimited Love as Ultimate Reality

Over a hamburger lunch one afternoon, Sir John Templeton said to me as he smiled broadly, "When I hear a choir, I'm inspired by the harmony of voices united. The same is true when I look at a pastoral scene, although I rarely take time to see much of nature. I admire a vibrant landscape blending flowers, trees, and grasses. That blend seems like singing voices to me." We had shared a few Chinese meals previously, and we mostly talked about finances and Unity writings of Charles Fillmore (cofounder with his wife, Myrtle, of the Unity movement).

So on this occasion, the "poetry" I heard in the above statement caused me to ask him to tell me more about what he believed. He was a many-faceted man, and the question was not answered in that brief lunch. As I recall, we met in early 1971 and became friends, and over the years I learned much more about what he believed.

When I was minister at Detroit (Michigan) Unity Temple, Sir John called me from New York to say he would like to take me to lunch the next day. He also told me he'd been listening to my radio program from Detroit in the Bahamas. I asked him, "Is this 'the' John Templeton?" His quick-witted reply: "No, my son is another one, and there may be others."

Among other phrases and fairly comprehensively full-blown ideas he discussed was the idea of Unlimited Love and Ultimate Reality, which fostered a brief conversation about Plato's *Symposium*. What I recall of Sir John's comment was, "This is another way in which love manifests a desire to have what we value most to last forever. As Plato explains in connection with the beauty, or appealing character, that arouses love, what is *most* forever is what is unchanging. And what is most unchanging is absolute truth and absolute reality. These 'absolutes' are simply what they are, and therefore involve no change at any time. Love, then, as a desire to possess forever, is really an experience of the desire for ultimate, absolute truth and reality."

Sir John quickly added that he admired Charles Fillmore's work along these lines.

Graciously, when speaking of Charles Fillmore's beliefs and Unity teachings, Sir John would often introduce a thought by saying, "As you well know," or "You know this better than I do, but here's what I understand Charles was saying." Sometimes I did know well, often I did not believe I knew better than Sir John did, and sometimes I was familiar with the writing he would mention, and very often, what he said about it gave me a whole new dimension to look at and ponder. May I add that when we met I was thirty-six, and without trying, he became another teacher of mine.

Sir John had long since given up the idea that God is a personality or is a manlike being, or even a man exalted far above human characteristics, indeed if he ever believed that. I had the feeling he may have resisted any anthropomorphic view of God as a child. He told me his mother gave him *Weekly Unity Magazine* to read when he was fourteen. Shortly afterward, Sir John began reading *Daily Word Magazine* (the world's first published daily inspirational magazine) and continued all his life. He was a voracious reader and had read all of Charles Fillmore's books and "legions" of his articles. In his own early books, in particular, Sir John quoted Mr. Fillmore quite liberally.

Sir John also had a strong interest in science, which paralleled Charles Fillmore's interest. Mr. Fillmore always contended that science and religion had no legitimate quarrels with each other, and Sir John agreed strenuously.

Sir John had two monistic points of view: he held the philosophical point of view that there is only one ultimate substance or principle, whether mind (idealism), matter (materialism), or some third thing that is the basis of both. His other monism was that God *is* everything expressing *as* everything. In other words, the "thing" that is the basis of both mind and matter is God as "First Cause," and Sir John said this enthusiastically. (Enthusiasm in the Greek, *enthousiasmos*, is "inspire" or "to be possessed of a god." As much as the words he used, the *way* he said it revealed how much he believed what he just said.)

In the same conversation he made it clear that he believed the big bang theory, and numerous other "big questions" and *some* of their answers.

The way these two monistic thoughts came together in virtually the same conversation was further discussed in a different conversation a few months later by referencing the way Mr. Fillmore defined love and Divine Love and what they have to do with Ultimate Love.

When Sir John and I discussed the subject of love, this time he quoted Charles Fillmore in *Christian Healing*, verbatim from memory: "Love in Divine Mind (another name for God or 'First Cause') [parenthetical phrase was Sir John's interpolation for emphasis] is the idea of universal unity. When expressed, love is the power that joins and binds in divine harmony the universe and everything in it." He commented that some probably think that's too impersonal a belief, but he said he was quite comfortable with it.

We talked a bit about Unselfish Love, Unlimited Love, and that it was called *agape* by the ancient Greeks to distinguish the divine (impersonal) love from earthly (personal) emotions. Unlimited love meant total constant love for every person with no exception. Here he disagreed with the Greek view and reiterated Charles Fillmore's view. He went on to say that he felt God's unlimited love for "all" creation, and not just for all people, may be the basic reality from which all else is only fleeting perceptions by humans and other transient creatures. All the great religions have taught us how to love and worship. It is always ready to make allowances, to trust, to hope, and to endure whatever comes. Most of what I've shared here is not in quotes because I took no notes during our talks, but once I was in my hotel room or on an airplane I wrote brief outline notes which helped me recall pretty well the nature of this series of talks. Later I fleshed them out a bit on my computer.

Personal love often does come to an end. Divine Love never comes to an end. Love seems to be the ideal and the dream, in some manner of expression, of every person. Could our souls have been conceived in the Creator's love, and could it be that love foreshadows our destiny? We seem more fulfilled when we are in a state of spiritual love and, somehow, emptied when the focus of our faculty of love moves elsewhere. Is it possible that love becomes the purpose of our existence? Although billions of words have been written about love in its many expressions, not one, or all, of them can fully capture the essence of love.

We talked at length as well about the fact that people who love life and the work of their heads, hearts, and hands are also likely to love all people generally, some people quite specifically, with a love of all creation generally demonstrated as caring for the well-being of all animal and plant life as well as caring for the life of the planet.

I close with this story about loving life and the work of one's head, heart, and hands:

After I resigned (to return to church ministry) in 2005 as president after twenty years of the Association of Unity Churches International (d/b/a Unity Worldwide Ministries), I accepted an interim assignment in Sheboygan, Wisconsin. While I was there Sir John's son, John Jr. (Jack), called me and told me that if I wanted to see his dad again that I should go to visit him in the hospital in Nassau as soon as possible. I thanked him for calling and expressed my regrets at the then not-so-surprising news and said to Jack, "I love that man" and Jack said, "I know you do. That's why I'm calling." I arrived in the Bahamas the next day in late morning.

During my flights on my way to visit him, I reviewed the many times we had visited. He came to my church in Detroit in 1971 (he was hearing my radio programs in Nassau) on a weekday and flew home that afternoon. He came there two more times, one of which was for a four-day Healing Retreat with numerous Unity ministers; Carl Simonton, famous oncology physician; and other health-care practitioners. Again for a similar retreat in San Francisco. Also, I had visited in his and Irene's home twice, and he attended several of our association's international conventions. And there were many other meetings because he invited me to become involved with the John Templeton Foundation(s) and get involved I did, several times a year for twenty-two years so far. However it was rare that we had more than a five-minute chat.

It was in this airplane review of some of those meetings that I remembered an experience I had at a meeting in Tennessee. First the reader needs to know that I grew up in a quite large "hugging" family. And Unity, the faith I found at age sixteen, is a "hugging movement." On one occasion of greeting Sir John in Tennessee, instead of offering to shake hands, I somewhat "automatically" opened my arms to hug him. He quickly crossed one hand over the other arm and took my hand to shake it. Had I been completely alert not to do what came "automatically," I would not have attempted to hug him.

Back on the ground, and with my head out of the sky reverie, I realized I had not asked Jack what hospital he was in, so from the airport I called Sir John's office to speak to Mary Walker. We greeted each other, and I asked where Sir John was. She said, "He's in his office. Would you like to speak to him?"

After recovering a bit from the surprise, I said, "Yes, please." I was surprised but knowing him pretty well I was not shocked that he would have gone to his office within a few hours of release from the hospital. What a pleasant surprise!

We exchanged greetings and he asked, "What brings you to Nassau, and will we have time for a meal?" My response: "Yes, in fact, seeing you is why I came." He suggested I take a cab and meet him at the Lyford Quay Club for lunch. I asked if I could have the cab come to pick him up at his office and we go to the club in the same car. His answer was, "Oh no, I drove to work this morning and I'll just meet you there." I couldn't resist laughing out loud, so he did, too. "Vintage" John Marks Templeton Sr.

Over a two-and-a-half-hour lunch we covered briefly many topics, and he reviewed the stock markets for me. Less seriously I chided him about leaving the hospital and going straight to the office. To which he responded, "I told you I never want to retire and I still don't."

Then he picked up the thread of the conversations of unlimited love as ultimate reality. Again, this is not a quote but a close approximation of what he said: "You know that I believe that unlimited love is the ultimate reality." I nodded that I knew that and said that I agreed with him. He also said, "I didn't always believe that. I used to think they were not related."

He continued, saying I know you know this (and I had to smile) but God is not "in" people nor the universe.

God "is" the First Cause and therefore expresses "as" people and "as" the universe (and maybe universes).

I had called for a cab and it had arrived. I walked with him to the parking lot and over to his car. We wished each other a bon voyage and shook hands. Sir John opened the door and started to step into the car, but he didn't. He turned around and came back to me and opened his arms for a hug.

That was the last time I saw him or spoke with him. He lived another sixteen months.

DAVID MYERS, PROFESSOR OF SOCIAL PSYCHOLOGY,
HOPE COLLEGE

Selfless Love as the Supreme Virtue

The germ of John Templeton's idea of *selfless love as a supreme virtue* surely was rooted in the soil of Winchester, Tennessee. It was there that I came to know him, over early Foundation gatherings on a family member's porch in nearby Monteagle, and then through conversations, some of which occurred as we together drove the streets of his boyhood Winchester, shared a long ride to the airport, or when he would take me aside to solicit my support of his ideas of "spiritual progress," "humility theology," and a science that would explore "laws of life."

John Templeton's roots in Winchester and Andrew Carnegie's in Scotland both were (like my own in Seattle) nurtured by a Presbyterian culture that drove into our souls the idea of *stewardship*. We do not exist for ourselves alone. All that we have is a gift from God. Any wealth we possess is entrusted into our care—our stewardship—for responsible management and disposition. "Like good stewards of the manifold grace of God, serve one another with whatever gift each of you has received." (1 Peter 4:10, NRSV)

This idea of stewardship, anchored deep in Sir John's soul, extended to his passion for purpose, as expressed both in his encouraging youth to formulate their own statements of purposeful life and in his supporting the extension of purpose into retirement. The U.S. Conference of Catholic Bishops likewise embraces this "stewardship of vocation: Each one of us—clergy, religious, lay person; married, single; adult, child—has a personal vocation. God intends each one of us to play a unique role in carrying out the divine plan. The challenge, then, is to understand our role—our vocation—and to respond generously to this call from God."

For those of us rooted in the culture of stewardship, self-indulgent greed was scorned. Those coming of age in Depression-era Winchester would be embarrassed to live in a palatial house that strutted their wealth. And strutting his wealth was not John Templeton. In contrast to other billionaires, he had but one nice but not ostentatious home. Until he faced health challenges at the end of his life, he flew coach. In one of our car rides, I once ventured to ask him what resources the Templeton Foundation might eventually have—"something like Foundation X?" I inquired (naming a $300 million foundation). He just smiled and replied, "More than that" (giving me no clue how many times more).

Sir John's sense of stewardship infused not only his self-giving philanthropy, his passion for purpose throughout life, and his modesty about his own wealth but also his overarching idea of unlimited love. If God is love, and if God's spirit is present in each of us, then we realize our spiritual essence when we seek others' well-being and engage in acts of unconditional care and service on their behalf. Our stewardship of wealth and vocation feeds and is fed by our embracing love as the ultimate reality.

The narcissistic "me generation" culture of John Templeton's later years embodied the antithesis of unlimited love. By contrast, humility, modesty, and purpose—all grounded in a keen sense of stewardship—defined the Templeton virtues. Ostentatiousness was anathema.

And so it was that, near his life's end, he focused $8 million–plus on establishing an Institute for Research on Unlimited Love, which he hoped would increase human understanding of Divine love and human altruistic love. Through this institute, grants to world-class researchers are supporting work on the mathematics, evolutionary biology, neuroscience, psychology, and sociology of the roots and fruits of unconditional love, as well as theological and philosophical analyses of agape love as a spiritual force.

"How little we know, how eager to learn" was Sir John's (and his foundation's) motto. Scientific inquiry or virtues such as love, he believed, helpfully supplement ancient Scriptures. This respect for science also reflects a Presbyterian heritage that has always valued education, that respects both biblical and natural revelation, and that considers itself "Reformed and ever-reforming." The theistic worldview of John Templeton assumed that (1) there is a God, and (2) it's not any one of us.

That being so, the surest belief we can hold is the conviction that some of our beliefs err, and that, in a spirit of humility, we can be open to God's continuing revelation through science and reason.

The stewardship of wealth and vocation, the idea of divine love embodied, and a humble openness to learning—these are among the big ideas, implanted in the cultural soil of rural Tennessee, that animated John Templeton to his life's end and that will animate his legacy into the future.

HAROLD G. KOENIG, MD, DUKE UNIVERSITY MEDICAL CENTER

Unlimited Love as Ultimate Reality

What does "Unlimited Love as Ultimate Reality" mean, why was it so important to Sir John, and how is it related to health and well-being? These are the questions I will try to address in this statement.

Unlimited Love is the type of love that God (the Ultimate Reality) has for people. Unlike human love, which has limits and conditions, Unlimited Love (or agape love) does not have such constraints. God loves us in an unlimited way that we cannot comprehend. No matter what we do or how often we fail to live up to the high expectations of our religious faith ("Be ye therefore perfect, even as your Father which is in heaven is perfect" [Matthew 5:48, KJV]), God still loves us and wants the very best for our lives. This is what the word "Unlimited" really means—an infinite capacity to give without receiving anything in return—and why it is so characteristic of God, who needs nothing from us and possesses everything to give.

We as humans are incapable of Unlimited Love. Our love typically involves a social exchange. You scratch my back, and I'll scratch yours. You show love to me, and I will show love to you. These are conditions that Unlimited Love does not have. Given our inability to love in this way, how can God expect us to do so? Is this not unreasonable? "Be ye therefore perfect, even as your Father which is in heaven is perfect," says Jesus. As Christians, then, we should strive to have this kind of love for others, no matter how difficult or impossible it may seem. Without a mark to shoot for or goal to strive after, we cannot make progress. Jesus provides us with the goal in Matthew 5:48 above and reinforces the point in John 13:34–35 (KJV): "A new commandment I give unto you, That ye love one another; *as I have loved you*, that ye also love one another. By this all men know that ye are my disciples, if ye have love one to another." Thus, Jesus tells us to love others like he (God) has loved us. There can be no mistake that Unlimited Love is God's kind of love and God wishes us to have it. Yes, we fail. But every day this is the goal and path set before us—one that Sir John reminded us of.

Importance of Unlimited Love to Sir John

Working on his newspaper (*Research News and Opportunities in Science & Theology*) and advising him on problems related to his aging, I came to know Sir John pretty well over a period of nearly ten years. The subject of Unlimited Love was really important to Sir John, so important that he wrote at least two books on the topic (Templeton 1999; 2000b). He thought that loving in this way was the secret to bringing humanity together; resolving the differences in religion, race, and nationality at the root of hatred and wars; and achieving real health and wholeness. The subject of Unlimited Love was at the core of his motivation and tremendous drive—that was evident in how he lived his life and in his writings.

The best source for understanding Sir John's perspective on Unlimited Love lies in his book *Possibilities* (2000a), where he lays out his philosophy of life and views toward science and theology, which he called the Humble Approach. Note what he says in the beginning of the book:

"This book begins to explore the possibility that developing a humble approach in theology, which encourages research and engages carefully with science, may be even more fruitful than endeavors to reinvigorate inherited systems of thought, whether they be polytheistic, deistic, theistic, monotheistic, pantheistic, panentheistic or even older concepts. Gradually each of us may learn to feel *unlimited love* for every person, with never any exception, and be grateful for an increasingly rich diversity of thought emanating from research and worship in every land." (2000a, 10, italics added)

One of Sir John's primary goals was helping people increase their love for others and, in fact, learn to radiate love. He wrote, "Why should we waste even one day? At the end of each day, can we say we have learned to radiate to all pure unlimited love and helped our neighbor to learn the joy of giving love? Brilliance of mind is not the same as beauty of character, but both yield helpful results." (2000a, 136)

This is what he sought for himself, and he was very successful at it—as anyone who knew him will attest to. Sir John discovered that in order to love others in this way, people had to get their eyes off of themselves. He wrote,

> "If we depend too much on the visible world, or trust our own ability too much or love ourselves excessively, we may never learn to radiate love. Instead we may radiate self-concern, egotism and arrogance. Ironically, egotism may not allow us to find our true selves. Can self-concern cause us to miss our true self of grace and generosity and thus tend to sever our links with eternal love?" (2000a, 27–28)

> "If we free ourselves of self-will and surrender to divine will, as is often taught, then we can become channels for divine love and wisdom to flow to others." (2000a, 28)

> "If we get rid of ego-centeredness, we can become clear channels for god's love and wisdom to flow through us, just as sunlight pours through an open window." (2000a, 34)

In Sir John's mind, Unlimited Love was not selfish or self-centered but rather was centered on the good of others and was connected with Divine purpose. He wrote,

"These sayings could well be called laws of love. If we radiate love, we receive joy, prosperity, happiness, peace and long life in return. But if we give love only to gain one of these rewards, then we have not understood love. Love professed in expectation of any reward is not authentic love. When we learn to radiate unselfish love, are we fulfilling divine purpose? Are we opening the door to heaven on earth for ourselves and others?" (2000a, 138).

Sir John also felt that Unlimited Love could be cultivated, and there were things that people could do that would increase their capacity for this kind of love. This was also connected with our love for God. He wrote,

"It cannot be repeated often enough that we can learn to radiate love, but first we must practice using loving words and loving thoughts. If we keep our minds filled with good thoughts of love, giving and thanksgiving, they may spill over into our words and deeds" (2000a, 146).

"When we say, 'I love you,' we mean that 'a little of god's love flows from me to you.' But, thereby, we do not love less, but more. For in flowing the quantity is magnified. God's love is infinite, and is directed equally to each person, but it seems to gain intensity when directed to those who need it the most. This is the wonder and mystery of it, that when we love God we get an enormous increase in the quantity of love flowing through us to others." (2000a, 159)

Rooted in Yet Not Limited to Christianity

Sir John did not restrict the source of Unlimited Love to the Christian God only, and felt that love was common to all religions. In order to make his point, he quotes Ella Wheeler Wilcox's poem, *All Roads That Lead to God Are Good*:

> A thousand creeds have come and gone
> But what is that to you or me?
> Creeds are but branches of a tree —
> The root of Love lives on and on.

He goes on to emphasize her point further by noting, "There is room for many branches on the tree. The lifesap of pure unlimited love lives on and on." (2000a, 39)

However, many of his ideas about Unlimited Love *did come* from Christian Scriptures, which he quoted liberally. Consider the following:

> We may especially reflect god if we create out of love. Thus, in the end, humility raises the morally challenging question of whether possibly the purpose of humans is to become helpers in the accelerating creativity expressed in the famous equation in theology, "God is love." (1 John 4:8: "He that loveth not knoweth not God; for God is love" [KJV]; Templeton 2000a, 88)

* * *

Without god as our source of inspiration, we are not likely to bring forth much good. We may never learn to radiate love as long as we love ourselves selfishly, for if we are characterized by self-concern, we radiate self-concern. "Jesus then said to his disciples, 'If anyone wishes to be a follower of mine, he must leave self behind; he must take up his cross and come with me. Whoever cares for his own safety is lost; but if a man will let himself be lost for my sake, he will find his true self.'" (Matthew 16:24–25 [NEB]; Templeton 2000a, 127)

Sir John immediately follows this citation with a quote from Emil Brunner:

Every human relationship which does not express love is abnormal. In Jesus Christ we are told that this love is the whole meaning of our life, and is also its foundation. Here the Creator reveals himself as the One who has created us in love, by love, for love. He reveals to us our true nature, and He gives it back to us.

Later, in his chapter on "Laws of Spiritual Growth," Sir John returns to quoting Christian scriptures (2000a, 160):

When Jesus was asked what is the greatest law, he said: Thou shalt love the Lord thy God with all thy heart, and with all thy soul, and with all thy mind. This is the first and greatest commandment. And the second is like unto it, Thou shalt love thy neighbor as thyself. On these two commandments hang all the law and the prophets. (Matthew 22:37–40, KJV)

And he followed this quote with long quotes from 1 John 4:7–21 and Luke 6:32–38. Sir John realized that loving God was a necessary prerequisite for humans to experience and express Unlimited Love for others. He wrote,

Love of god comes first and makes it easier to love in other ways. If we want our enemy to see only our good qualities and not our flaws, then should we not set the example by looking first for the good qualities of the other? That person, too, is a child of god. God loves us both even though neither of us is yet perfect. (2000a, 164)

He had also earlier questioned whether the pantheistic view of God could produce the kind of love that could be experienced in a relationship with an intensely personal God, noting that,

Traditional pantheism can serve a useful purpose in suggesting the co-terminacy of spirit and matter and a personal relationship between the creator and creation. But it may not be compatible with the Christian concept of a personal god vastly greater than material things and who loves all of us and numbers the hairs of our heads. (2000a, 86)

The Practical Consequences of Unlimited Love

Sir John clearly saw more than just a theoretical concept in Unlimited Love, and understood this in terms of practical consequences, both in terms of providing guidance on how to live and spend our resources and in terms of physical and emotional health. He wrote,

Mother Teresa formed a new order of religious women who have lived among and helped the poorest of the poor in India and many other nations to develop fruitful joyful lives through divine love. Public as well as private charitable organizations could follow her example and methods for providing human services and love to the outcasts of our age. (2000a, 45)

People pursuing the humble approach express their love in charity to prevent or to alleviate suffering and to elevate the recipients of their love. Feeding the hungry, caring for the sick, clothing the naked are all compassionate in the short term; but in the long term, the real charity is to help the poor learn the spiritual traits that lead to progress, productivity, prosperity, dignity and happiness. Bestowing technology and know-how on people in poor nations is a blessing; but the lasting blessing may be people who can radiate love and joy as they research and teach the basic spiritual realities, which then lead to progress, improved skills, spiritual wealth and also material prosperity. (2000a, 129)

Sir John realized that Unlimited Love had many spiritual and psychological benefits for the individual who loved in this way, noting,

The Christian religion speaks often about the way in which the human experience on earth is to function as preparation. To progress spiritually can be to increase our love of god, or understanding of god, and love for his children. (2000a, 132)

Who are the happiest people you have ever met? If we were to write down the names of ten persons who continually bubble over with happiness, would we probably find that most are men and women who radiate love for everyone around them?" (2000a, 137)

Sir John strived to love others in this way, and this was evident from the way he treated people. This, I think, was one of his secrets to the tremendous success that he had in the business world. During the time that I knew him, Sir John always expressed a sincere interest in other people (from cabdrivers to fledgling researchers to leaders of state), asking them about themselves and what really excited them. In many respects he was providing a role model that he hoped others would follow.

Sir John also realized human limitations, emphasizing that Unlimited Love was an ideal to be striven toward, but unlikely to be fully realized here on earth. However, there were things that could be done to help us move closer to that goal. He wrote,

If perfection may be beyond human reach, each human only can study and strive toward such divine love. Even the saints need to work daily to maintain continuous overflowing love for friends and foe alike. But the more we love, the easier it becomes to love even more. Love given multiplies. Love hoarded disappears. (2000a, 137)

Unlimited Love was a major focus of Sir John's life and writings. The evidence for this lies in the life he lived and in the book where he most clearly expresses his vision and legacy (a text that he revised over and over again during the last twenty-five years of his life). Sir John's views were deeply rooted within the Christian tradition, from which he could never completely separate himself. His ideas later broadened to value the wisdom expressed in other religious traditions, including the tradition of science, which he believed could discover spiritual truths that would make the ancient great religions more relevant to people today. One of the most important truths and laws of life to Sir John was Unlimited Love. Let us hope that science can help us document the spiritual, psychological, and physical benefits of this queen of all virtues, benefits that Scriptures have emphasized for over two thousand years. As Sir John emphasized, demonstrated, and encouraged others to do, Unlimited Love is as relevant to the modern world as it was to the ancient one, maybe even more. Some things never change.

Chapter 11

How Did You Experience Sir John as a Person Who Tried to Live a Life of Love?

DR. JOSEPHINE "PINA" TEMPLETON,
WIFE OF JOHN M. TEMPLETON JR., MD

"Papa Templeton"

Dear Stephen,

Love was the fuel that revved up Sir John's engine, but perhaps you already know that since you are probably the person with whom Sir John might have expressed more clearly what love meant for him as a general principle as well as what it meant to him as an individual.

What did he mean by love, and why was it so important to him to start his shareholder meetings and many other meetings with prayer followed by the salutation, "Welcome, I love you all"? He lived his life by the essence of love. We see this in the love he had for his wife and children, how thoughtful he was of them, the concern that he had for their well-being, his wish that his children would know God and that they would be good stewards of the blessings they had received, including making their own way into the world. The regard that he had for his brother Harvey (it was very endearing to see them sit together on Becky and Handly's patio as they reminisced of their younger days, as well as reflecting on the issues of the day); the interest he showed in the endeavors of his extended family; the respect he had for his investors (for he thought of himself as the steward of their savings and livelihood), the people who entrusted their money to him and often their eventual livelihood; to the people who worked in his offices and later his foundation; and finally to his house staff that cared for him. Why was love so important to him? Was he on a quest to understand its meaning and its power? He certainly put in a lot of thought, time, and fortune in understanding the highest forms of love. What was his motivation? His mother, Vella, certainly had a great impact on shaping young John Templeton, but there had to be a deeper reason for his search into the meaning of love. We know that as a young man, Sir John did teach Sunday Bible school. I think that if and when he taught 1 Corinthians the passage on love truly had to have made an impression on young John:

> If I speak in the tongues of men or of angels, but do not have love, I am only a resounding gong or a clanging cymbal. If I have the gift of prophecy and can fathom all mysteries and all knowledge, and if I have a faith that can move mountains, but do not have love, I am nothing. If I give all I possess to the poor and give over my body to hardship that I may boast, but do not have love, I gain nothing. Love is patient, love is kind. It does not envy, it does not boast, it is not proud. It does not dishonor others, it is not self-seeking, it is not easily angered, it keeps no record of wrongs. Love does not delight in evil but rejoices with the truth. It always protects, always trusts, always hopes, always perseveres.

And now, Stephen, as I continue, these will be only my reflections as I had the privilege to have shared in part of the life of the man who was my husband's father. When Jack spoke to his dad of the girl he wished to marry, he listened intently and then he had three questions: "Is she thrifty, is she a Christian, and do you love her?" And with that, Stephen, I began my life in the Templeton family and with Papa Templeton, the man whom I grew to love and respect and who became more than a father-in-law. This was to become more evident to me, for after my own father died, Papa became my father in situ. Papa was not a perfect man, but he was a loving man. Why? Because he was sure that there was something special in every person he met.

As life progressed, my fascination in learning the mind and heart of John Templeton continued to grow. Love for him was the great motivator, what made his glass half full, what made him get up in the morning and be joyous. This is where you come in, Stephen. I think he saw in you, a willing searcher, a willing partner in his quest to understand the many facets of love. God is love, I think Sir John truly believed that, but being the eager student, he wanted to learn as much as possible of the meaning of love not only for himself but for others as well. And that in itself is the quality of a loving man.

In William Proctor's book *The Templeton Touch* we find that reason for the Templeton Prize. Why did he establish a prize, which he first called a Prize for Progress in Religion? As is described in *The Templeton Touch*, Sir John explains that while he had made a certain progress in his spiritual life, his "friends" who were very successful in their professional and social lives did not seem to have made the same progress in their spiritual lives. Therefore, the prize was established to first assure that the work of these special individuals—who did traverse through the path of spiritual progress—could be made available to a wider audience but also gave his friends the opportunity to go further in their own spiritual journey. Is that not what love is? To be concerned that others may benefit of one's own experiences.

As you probably know, when Jack and I met we first became intellectual friends. He had a girlfriend, and I was to return to Italy after internship. This friendship gave us much more freedom (I think) to discover each other. It was during one of those dinners while on call that Jack told me that his dad, after analyzing Jack's work at school, had come to the conclusion that yes he would be willing to support his son's further education. At first this attitude may seem cold and calculating. Instead, I think it demonstrates a father who followed his child very closely while he tried to decide what was best for his son.

As you probably know, Jack's mom, Judith, and dad took their first vacation after the first company was established and was solid enough that they could take a vacation. It was during this vacation that Jack's mom died secondary to what we can deduce was an epidural bleed as result of a scooter injury. Sir John had to come home and tell his children that Mommy was not coming home. I can only imagine the anguish that this man had to have had experienced, having lost his own mother six months before. Jack told me that though they had not owned a television (probably because Papa did not think that it was of any value), his dad, on returning home, quickly went out and bought a TV for his children. This may not seem much but it is the love that this man (who very likely was trying to make sense of what had happened) had for his children that he quickly thought of something, anything, that could make the loss easier.

I can see the love he had for his children in some letters that I found (not put in sequence and sporadically found in his papers): letters to camp directors, physicians, to the Barbizon modeling school (Candy must have shown interest or he thought that practicing modeling would give her self-confidence). A man who follows his children so closely is a loving father.

When Jack and I became seriously involved (after he pierced my ears and gave me a topaz ring that he had bought years before, a ring that was going to be for the girl he married), it was time to talk with my dad, which he did; but also he would have to talk with his dad (although Jack will never admit it, he had to be nervous, for I was not American born, I was raised Catholic and certainly not a girl from their circle of friends). His father listened intently and then remarked (I paraphrase.) yes, she seems to be a nice girl but let me ask you, "Is she thrifty? Is she a Christian? And do you love her?" One can chuckle at this, but to me it was a father who, maybe surprised by the news, was trying to evaluate whether this girl would be a good companion for his son, and that is a loving father. Jack then returned with a diamond wedding band and engagement ring that his dad had given to Jack's mom. (I always thought that it should have been given to his daughter but he must have had his reasons.) I think that Jack's dad kept these two rings as a sign of love (what I think was his great love) for Judith, Jack's mom, and as a sign of love he gave them to his son.

Jack's dad readily accepted me. He did scrutinize me for a year (very appropriately). Maybe he wanted to make sure that I was truly thrifty. He always showed interest in what I was doing; that I worked; but also how we raised our children, never demanding yet observing.

While in training, Jack developed hepatitis, which could be traced to a patient he had cared for. I was on rotation in the emergency room, and when my day was finished I would visit Jack. There was always a daily call to check on his son and whether he (his dad) should get the best experts in the world. We were at Medical College of Virginia and had more than the experts, but then again he showed what Love is. He showed the same concern for me when in 2002 I developed sepsis after our trip to Spain. There was that daily call to make sure that everything was going well and that we had the right doctors.

Papa had to have surgery for glaucoma. I am not sure I remember why I went to take care of him rather than Jack (Jack must have been on call or something like that). I picked Papa up at the airport in Miami with a rental car to drive him to the clinic. He was appreciative but also told me that it was not necessary; that he could manage. Well, we agreed that I would drive him to the clinic and then he could decide. (On the other hand, no matter what he had told me, I was going to wait for him.) I went up to the presurgical area to see if he was ready to go in to the O.R., only to find when I got there that the nurse was waiting for me. Papa had told her that he was not going to go to surgery until he could see his daughter (I had to tell her that I was his daughter-in-law). In fact, he would often introduce me as his daughter (even to Queen Elizabeth), which was very confusing at times because he had already a wonderful and loving daughter. Is that not Love, that he would accept me as such?

What was interesting to me was a comment he once made as we were talking during a walk together in Jamestown, North Dakota. I don't fully remember our conversation, but at one point I remember telling Papa how much his children loved him. He stopped walking, which startled me, and asked me if I really thought so. I found that interesting that a man who could love so much was not sure if he was loved.

Along the way in my married life I had the opportunity of meeting a number of people who had benefited from Sir John's concern for them, even though he did not know them well. At the last World's Fair in Vancouver we were met and driven by two stockbrokers who felt indebted to Sir John. One gentleman went on to tell me his story. A number of years before, he had lost his wife to cancer and became very despondent. Even though he loved his children, he slowly became addicted to alcohol— almost to the point of destroying his career and his relationship with his children. It was during this period that this gentleman was invited to a meeting where Sir John was one of the main speakers. At the end of the evening the gentleman asked Papa if he could have a few minutes of his time. Papa told him (I paraphrase) that he was busy at the moment but could he have his hotel room number and he would see him later. The man waited and waited; it was nearly midnight, at which point there was a knock at the door and Sir John was there to talk. They apparently talked for hours. Whatever Papa said gave this man enough confidence to turn his life back to a life of accomplishments and productivity. Why would Papa do this unless he loved?

And so, Stephen, these are a few of my reflections of a man who was forever a seeker but whose life was founded on or around love and Love was very likely the queen of all virtues.

HEATHER TEMPLETON DILL, DAUGHTER OF
JOHN M. TEMPLETON JR., MD,
AND JOSEPHINE TEMPLETON, MD

Grandpadaddy's Affection

I never talked to my grandfather about Unlimited Love. That is my fault. I was always a bit intimidated to speak with him and frankly too immature or too interested in my own life to engage him in any significant way. It is one of my greatest regrets. I have learned so much more about him after he passed away and have come to value his vision for increasing knowledge of spiritual realities. I wish I had taken more time to know him when he was alive.

But I am still a member of the Templeton family, and Sir John's vision, his humble way of life, and his joy trickled down to each succeeding generation. We were taught to value hard work. We were taught to live simply. Sir John never flew first class, for example. His home, although beautiful, was modest in comparison to what he could afford. Moreover, we learned that serving others was the mark of a life lived well. He often spoke of his career in the mutual fund business as a service to others. He used his expertise to help others save their money and preserve wealth so that they might be better stewards of their resources. I think he found a great deal of meaning in that vocation.

He found even greater purpose in the mission of the John Templeton Foundation, which he established in 1987. His success in business enabled him to seek ways to save religion from obsolescence, increase spiritual knowledge through scientific research, and encourage more people to live useful and purposeful lives. I remember my grandfather saying on more than one occasion that although he sold his business, he was busier than ever. And he was more joyful than ever. I can't remember exactly what he said, but I can still see his smiling face and hear his voice with that faint southern drawl.

While it is only recently that I have learned how much Sir John thought about love and the extent to which he was interested in divine love, I grew up with a grandfather who greeted us each morning with a big hug and a warm "I love you." My own immediate family did not express our love so effusively; Grandpadaddy's affection and uninhibited use of the word *love* made an impression. I don't remember that I ever returned the greeting in the same way. I think I was too shy to do so. But the very fact that I couldn't get those three simple words out of my mouth demonstrates how much my grandfather valued love. I do remember sharing this story about Sir John's daily greetings with the staff at the foundation after Sir John had passed away. Clio Malin later told me that Sir John would also hug her and say, "I love you." In some ways, she hardly knew him and he hardly knew her; but his expression of love extended far beyond his own family.

I don't think Sir John always articulated love in this way. I never asked my grandfather about how the loss of his mother and his wife within three months of each other affected him. My aunt and Sir John's only daughter, Ann "Candy" Zimmerman, told me that she taught my grandfather to say, "I love you." He had not always expressed love so openly and that is no surprise given the hardships he faced in life. At some point in her life, Ann decided to reach out to her father with words by saying, "I love you," and with actions by holding Sir John's hand or giving him a hug. Over time, as she described it, he too adopted these practices. In some ways, his ebullient display of affection and love developed in response to those who had demonstrated love toward him. In other ways, I think his daily effort to say, "I love you," focused his thoughts. In *Wisdom from World Religions,* he wrote, "Love, as with any other spiritual virtue, doesn't simply fall into our life as manna from heaven. Like an inquiring mind, it needs to be cultivated." The act of saying, "I love you," and sharing an embrace was in part his way of cultivating love within himself and sharing that love with others.

And, indeed, although I never developed an intimate relationship with my grandfather, I never doubted that he loved me. He took a keen interest in my activities. He came to my high school graduation; that was amazing to me. He also attended my wedding. When my sister and I were very young, Mom had to leave us in the Bahamas for a few weeks because she had no other option for childcare. My grandfather visited us every day on his way to work and on his way home. He also gave a great deal of thought to our future life partner. My sister and I laugh about how much he considered the matter of our finding a good man to marry. He suggested that we apply to one of the state tech schools such as Georgia Tech or Virginia Tech because the male-female ratio was distinctly in our favor. When I ended a two-year relationship with one beau, my grandfather invited me to spend the summer at his home on Fisher's Island to meet the eligible young men who summered there. I could give more examples, and while I still chuckle about his interest in my dating prospects, I always felt that his concern about this spoke to his deep affection for me.

In June, Mary Mazzio interviewed me for a documentary about Sir John. I shared a few stories about my grandfather's interest in introducing me to marriageable young men. She asked me if I ever felt that his queries about my romantic involvements were related to wealth and inheritance. I could not understand the premise of the question. It had *never* occurred to me that my grandfather's interest in finding me a suitable mate might have to do with my material well-being. I always thought that he wanted me to marry well so that I would know the joy of sharing life with one very special person. Perhaps I am wrong about that. But I think my negative reaction to Mary's question and my assumption that Grandpadaddy only wanted me to be happy say more about the kind of love that directed his life. Although he could not spend time with us the way many grandparents do, he thought about us, he provided for us, and he always welcomed us with open arms and an open heart.

For Grandpadaddy, love was "total constant love for every person with no exception." At the end of one chapter in *Why Are We Created?*, he writes, "God loves you and so do I." These are not idle words for Sir John, and I wish I had realized how much he welcomed the input of others and how much he would have happily entertained my questions. He was always positive and never threatening; he smiled often and listened attentively. These are marks of a man who focused on sharing love with others and found joy doing so. And this is the legacy I seek for my own life as well as the life of my children.

JENNIFER TEMPLETON SIMPSON, DAUGHTER OF
JOHN M. TEMPLETON JR., MD,
AND JOSEPHINE TEMPLETON, MD

Grandpadaddy and the Love of Neighbor

I did not grow up living near either set of my grandparents. And I wish I had. I am so grateful that my children do live near all four of their grandparents, who are all still vibrant and full of health. The love of a grandparent for a grandchild is special and different than the love a parent has for a child. It is one that I know I do not fully understand since I am not yet there, but I imagine it is a love that is filled with all of the emotion a parent feels for a child with the incredible addition of watching a child take on the role of a parent. I do hope I get the privilege to be grandparent someday.

Although I missed the opportunity to be close emotionally to my grandparents because of physical distance, I did feel so blessed to be related to people I considered so incredibly special. My mother's parents immigrated to the United States from the island of Capri, an island so beautiful that my Nonna promised me that if I ever mentioned I was from there during a job interview, I would be most certainly hired. My Nonno was a lieutenant commander in the Italian Navy and an accomplished engineer who would have fixed any of my broken toys—therefore making him a hero in any child's eyes. My Italian grandparents had moved back to Italy by the time I was born, and my Nonno died when I was six years old. But their stories are part of my story and something that helps to define my identity.

I never had the chance to meet my paternal grandmother, who died when my father was around ten or eleven years old. From what I hear she was every bit the mover and shaker that my grandfather proved to be and they were a perfect match. Grandparents can often seem like heroes and certainly role models in the eyes of their grandchildren. In my experience, my paternal grandfather, Grandpadaddy, was not only a hero and role model to his grandchildren but to many people throughout the world.

I rarely ever thought about my grandfather's fame or his influence in the financial world, so it would genuinely surprise me when strangers would know his name. One of my earliest memories of realizing that people besides my family knew my grandfather came when I was about eight years old. My sister and I were eating breakfast before school, and the radio was on. The report was about a critical day during some financial crisis, probably the savings and loan. The reporter said that he was very unsure of how the day would turn out because neither God nor John Templeton had returned his call. My sister and I immediately turned and looked at each other, shocked and amused. I enjoy sharing that memory, but it is far from my favorite in terms of the memories that I hold closest to my heart. The memories that mean the most to me are the ones when I truly learned the level of love my grandfather could feel for others.

When I was around ten years old, my family and I went to Vancouver for an exciting trip to the World's Fair. When we arrived we were greeted by someone who knew us and was waiting to give us a ride to our hotel. I always enjoy being met at the airport, but what made this event memorable was overhearing the man driving tell my parents that my grandfather had saved his life. The man was a widower whose wife had died of cancer and had left him with young children. He had turned to alcohol to deal with his grief and stress. Alcohol soon became the predominant problem, and he was on the brink of losing his job and his family when he met my grandfather at a meeting. After hearing my grandfather present, the man, who was a stockbroker, waited in line to speak with him. My grandfather had to leave by the time the man was at the head of the line, but he asked the man for his hotel room number and said that he would come by after the meeting. The man returned to his room and waited until midnight, and just when he was about to give up and go to bed, thinking that this was another disappointing moment in his life, my grandfather arrived, sat down with the man to talk, and stayed until two in the morning. My grandfather did not give the man any money nor did he offer him a job, but whatever he did say to this man that night enabled the man to turn his life around.

I remember even at a young age being so amazed that just the words that my grandfather said were enough to help someone face things as difficult as grief and addiction. I also remember the expressions on my parents' faces as they were also amazed and filled with pride and admiration for my grandfather and the impact that he could have on people. The story told by this truly grateful man to my family is the first time I understood why my grandfather was a special man. It was not because he had made a lot of money or was internationally famous but because he wanted to and did strive for higher purposes. My grandfather wanted to live for more than the riches of life; he wanted to know, understand, and practice the richness of the spirit.

Although I did not see my grandfather very often, I was always sure that he loved me. I could be sure of this because every time I saw him he greeted me with a bone-crushing hug and a giant "I love you." When I say bone-crushing I mean every aspect of it. In my teens I finally admitted to my father that I would prepare myself when I knew he was approaching the door by allowing my body to go a little limp in order to be able to enjoy the enormous bear hug coming my way without feeling the bits of pain that came along with it. The hug was coupled with a loud "*I Love You.*" My grandfather was very free with the words "I love you" for me and my sister in what he wrote to us and what he said. I learned that you can never say "I love you" too much. He did not shy away from expressing love no matter the audience. Truly expressing love to others is harder than it may sound, and for my grandfather I believe this is something he only learned to do over time; it was not part of his initial natural way. Somewhere in his progression in life he learned, came to accept, and wholeheartedly promoted the importance of love.

From time to time, I have been asked what my grandfather shared with me either about his views on finances or his hopes for his foundation. I honestly do not have any or at least no significant memories of my grandfather wanting to share his take on the economy. I have some memories of my grandfather explaining why he started the foundation along with several interesting stories about the challenges of communicating to the world his intentions for the work the foundation would fund. In reality these conversations with my grandfather were few because what he always seemed to want to talk to me about was one thing: Love. It seemed very important to my grandfather that I would find someone to love who would love me. He seemed very keen on telling me stories of how much he had loved my grandmother. I, of course, loved hearing stories about the grandmother I never had the chance to meet, but I also loved these stories because I thought it was amazing that out of anything he could choose to talk to me about it was always this. Because he so loved this topic, I did end up hearing some of the stories more than once or twice, but I never minded. I loved seeing the spark in his eye when he would recall just how much he loved filling my grandmother's dance card and just what a catch he believed she was.

I most often heard my grandfather's stories about my grandmother while I was visiting him during spring break. He would work most of the time I was there but always met me for lunch. One day, I thought we were about to embark on a topic change when he started out by asking what colleges I was considering. I am not sure of my response, but I will never forget his. He asked if I had thought about Georgia Tech. I responded no, very curious of why he suggested a college so far away from my home and full of majors not with my interests. He informed me that the ratio of women to men there was 1 to 4 and that he had high esteem for the students who attended this school. He thought it would be a good place for me to search for a good match.

One could think less of my grandfather for suggesting that my higher education choices should include considerations of searching for a marriage partner, but I knew that my educational and professional success was important to my grandfather. My grandmother was an accomplished professional in the advertising field and at one point the family breadwinner. What I took away from my conversation with my grandfather was that he was not worried about whether I would focus on school, but he was concerned that I would not focus enough on finding love. Although he never said this, I think he thought that one of his greatest achievements was his partnership with my grandmother.

I did not end up going to Georgia Tech. I went to Boston University where the ratio was not at all in favor of more men to women. While home from a visit from school I found a letter from my grandfather to my father that was accidently left in my room. It was a handwritten letter probably about three to four pages in length. My grandfather was suggesting to my father that he throw me a type of coming-out party to which any and all potential suitors would be invited. My father could employ the dance card system so that I could be sure to meet all of these potential life partners. I have to admit that my initial reaction was immediate and extreme fear that this plan was already under way. After, I found my father and had him swear to me that in no shape or form was he going to host such a party, and my fear was replaced with a warm feeling of love for my grandfather. He was such a busy man with many demands on his time, but somewhere in his day he was finding time to be concerned about my love life.

I did find the love that my grandfather wanted for me in my best friend. We are the proud parents of two wonderful children, which fills us with the amazing love a parent has for a child. And I get to share these children with their grandparents and aunts and uncles who add even more love. I think that my grandfather would feel so satisfied with these connections that are based on love. I think my grandfather would not want us to take all of this love for granted, but he would want us to think about it each and every day. He would want us to wonder why this emotion is so important. He would want us to wonder how many types of love there are in the world and the power of each of these individual types of love. He would want us to be ever searching about what it means to love and to be loved and how we can love better. Out of everything that one could expect that my grandfather would want for me to care about, I know that what he wanted me to focus on first and foremost was love. I say this without doubt. I think he believed in the power of love to do so many things, more than we could expect an emotion to do. I hope to pass on this legacy to his future and continuing generations.

PAMELA THOMPSON, DIRECTOR OF
COMMUNICATIONS (EMERITUS), THE JOHN
TEMPLETON FOUNDATION

"A Little Bit of God's Love Flows from Me to You"

Perfectly turned out in a dark green linen jacket over a herringbone white shirt and crested Yale College tie, Sir John Templeton rested his arm across the back of the turquoise and coral chintz-covered sofa at the Lyford Key Club. It was a magnificent evening; small waves churned into the sand beach as the enormous sun seemed to drop into the ocean. Four petite women adorned in sparkling diamonds were hanging on John's every word as he explained some of the scientific research his foundation was funding on spiritual subjects such as love. Sir John Templeton used to say that one could not give away enough love; that the more love we give away, the more we have left; and that "love hoarded dwindles and love given grows." When he left a gathering of friends and colleagues, he would always say, "I love you," meaning that "a little bit of God's love flows from me to you."

John loved his older brother, Harvey. As children they spent most of their time together in Winchester, Tennessee. They grew up living next door to their grandparents and had free run of their large neighborhood by the age of six. Everyone knew one another, and your word was your bond. As the brothers grew older, married, and had children, the families remained close. John and his wife, Dudley, and children, Jack, Anne, and Chris, lived in Englewood, New Jersey. Sadly, Dudley was killed in a motor scooter accident while the family was on vacation in Bermuda. For several summers after that accident, the children spent their summers in Winchester, staying with their uncle Harvey and aunt Jewel and cousins, Jill, Harvey, Handley, Avery, and Ann. Some of their favorite memories filled with hilarious episodes were about their Florida trips with all of the young cousins in Aunt Jewel's black hearse, called the Queen Mary.

It was not long before John married a second time. Petite and elegant,

Irene Butler had two handsome children, Malcolm and Wendy, and with his three, they moved into his Englewood house and began spending their summers on Fisher's Island. In 1968 John and Irene decided to move their permanent home to Lyford Cay, Nassau, in the Bahamas. John was already exploring ways to spend time for spiritual progress. He said, "I had spent my early career helping people improve their personal finances, but helping them to grow spiritually began to seem so much more important." John soon became a British citizen and established a program for spiritual training for Bahamians interested in the Christian ministry, and he provided several fellowships for study in the Bahamas and at Princeton Theological Seminary, Princeton, New Jersey. John's spiritual commitment was growing. Robert Herrmann wrote in his biography that "Sir John Templeton never overlooked the importance of prayer." He quotes John as saying, "We start all of our meetings with prayer, prayer helps you to think more clearly. We ourselves seem to be a recent creation of God and a little part of God. If we realize this and try to bring ourselves into harmony with God, whatever you do in life, you should open with prayer, and that prayer should be that God will use you as a clear channel for His wisdom and His love."

Four years later, John established the Templeton Prize with much excitement and enthusiasm, which was not surprising since his desire to love and help everyone had been emblematic of John's behavior since early childhood. While his investment success was increasing, his desire to showcase some of the marvelous new things going on in religion was also growing. He said, "There were new churches being formed, new schools of thought arising, new books being written, and new denominations appearing. And I thought, how wonderful it would be if my friends could hear about these things and read about them. They couldn't help but be uplifted and inspired if they could just be informed about what was happening." With guidance from longtime associates Dr. James McCord, president of Princeton Theological Seminary, and Lord Thurlow, the governor of the Bahamas, John decided to offer a prize for progress in religion—all types of religion. He organized a system of rotating nine judges on three-year terms with half from the five major religions and half not, to promote "the idea of being receptive to new ideas." John also wanted a famous person to present the prize to bring attention to the prize and the recipient. His Royal Highness the Duke of Edinburgh accepted the long-standing invitation and presented the first prize in 1973 to Mother Teresa of Calcutta. The judges had chosen her to celebrate her ability to love the unlovable. She received the Templeton Prize five years before she received the Nobel Peace Prize.

In 1987 John Templeton was knighted by Queen Elizabeth II, and the same year he created the John Templeton Foundation. John's family, childhood friends, and lifelong colleagues became the nucleus of his foundation, where he believed scientific research could encourage progress in obtaining new spiritual information and unlock the potential of unlimited love and other important spiritual principles to promote progress in religion. Sir John hoped that religion would become just as progressive as medicine or astronomy. He thought, "Religion should be just as exciting as any other field. Probably in the long run, the manpower and money we invest in discovering more about God should approach what goes into science." For the past twenty-five years, Sir John and his foundation have been trying one thing after another to do something that increases our knowledge of God, God's purposes, or God's love. Sir John believed, "No one can know the total truth, everything about God, or the intricacies of our beautiful universe, but the rigorous method can start movement in the right direction." Scientific research, he said, "is part of God revealing himself and God reveals himself to those who seek to learn more about the purposes, the reality, or the infinity of God than we ever could imagine."

Celebrating Sir John's eightieth birthday on November 29, 1992, Sir John was honored with a Festschrift. Many articles were collected from some of his most adoring supporters. A Foreword was written by His Royal Highness, the Duke of Edinburgh. Prince Philip's words alert us to Sir John's extraordinary place of both humility and love in his relationships with treasured members of his family, friends, and lifelong colleagues:

Dear Sir John,

The world is full of experts and specialists, but it is given to few to be pluralists; those exceptional people who not only comprehend the great issues, but whose questing minds can go straight to the heart of the matter. Your highly successful business career and the wide scope of your interests bear witness to both the clarity and the originality of your thoughts. You have backed this up with a truly magnificent generosity to the many causes close to your heart.

It is not the passionate revolutionaries who improve the world; the only genuine progress is made by those who have a clearer vision and a greater ability to see through the tangled web of humbug and convention. Your great contribution to human civilization has been to encourage people to concentrate on the things that really matter and to appraise both conventional as well as unconventional ideas for their true worth.

John Marks Templeton was born one hundred years ago in Winchester, Tennessee. His talent as a money manager is legendary, and his foundation has focused on an undertaking that has made Winchester's culture known to the world. In June 2012, more than three thousand area residents attended Dr. John M. Templeton Jr. and the John Templeton Foundation's Spirit on the Square, a festival honoring Franklin County/Winchester residents and celebrating the values Sir John embraced. Spirit on the Square featured award-winning country music stars and other outstanding performers. However, the centerpiece of the event was the competition among twenty-five nonprofit organizations with each having displays around the town square. The contest marked the twenty-fifth anniversary of Sir John's *Laws of Life* Essay Contest by asking each nonprofit to showcase its mission along with one of Sir John's core virtues, such as love, humility, and progress, which had a profound influence throughout his childhood and early adult life. Through his foundation, Sir John was able to give away much love to residents of his hometown.

I met the eighty-four-year-old philanthropist while he was holding interviews to hire a director of communications for his new foundation. As I walked off of the elevator in the Union League Club of Philadelphia, Sir John Templeton greeted me with an enormous smile and said, "What should I call you, my great, great- or grand, grand-niece?" Shocked I replied, "What do you mean, Sir John?" "Well," he said, "your father-in-law was not only in the same class with me at Yale, but since Wirt Thompson was an international champion pole vaulter and I was a Rhodes scholar, we were in the same fraternity." My nervous feelings evaporated as this kind man led me through a demanding interview, and when we were finished, Sir John and I shared an urgency to promote the results of his extraordinary scientific research on spiritual subjects to increase spiritual information.

Until Sir John died in 2008, we worked together to brand the mission of the foundation and each of its grantmaking areas, publicize high-quality scientific research results through top scientific journals and conferences, and find outlets to showcase his commitment to encourage enthusiasm for applying scientific methods to the discovery of over a hundredfold more about spiritual realities that are not tangible or physical. Sir John sent me faxes with brilliant ideas and suggestions almost every day. Our task was gargantuan, but his amazing energy, openness, conscientiousness, and intellect kept moving us along. His vision was to encourage scientific research to procure empirical data about spiritual realities such as love, purpose, creativity, thanksgiving, prayer, humility, praise, thrift, compassion, truthfulness, giving, and worship. He was also committed to science exploring the big questions. In the human sciences and in character education, he supported programs, competitions, publications, and studies to promote the exploration of creativity, positive values, and purpose across the lifespan. He supported freedom and free enterprise education through programs that encouraged free-market principles. Sir John noticed scientists, theologians, and scholars who could envision ways to unlock new information about these initiatives. These colleagues became very dear to Sir John, with many of them becoming members of his board of advisors because of their similar attraction to groundbreaking discoveries and practices.

BARNABY MARSH, VICE PRESIDENT FOR STRATEGIC
INITIATIVES, THE JOHN TEMPLETON FOUNDATION

Sir John Templeton and the Mysteries of Unlimited Love

"The mystery of love is that when it is shared, it grows, and further, the more you share, the more that you have left," was one of the favorite sayings of Sir John Templeton. In my personal one-to-one meetings with Sir John, he would often use this saying and the sense of wonder it instantly created as a launching point for an intense discussion to follow that would frequently entail a lot more questions than answers.

"Could we love better, purer, and more effectively?" "Is it impossible to love too much?" and "Is love the ultimate reality?" were only a few of the questions we pondered and pondered. We explored possibilities in detail, and John would add a lot by reflecting on sayings of prophets, sages, and even ordinary people. Were there universal principles or specific insights that could sharpen love's broad richness? He would discuss the limits of the human mind, and how this might influence what we are able to sense, and what may be beyond what we could sense. We discussed love in the context of the individual, and in the context of the connectedness of souls in the world and between generations and times. We discussed love's transformative powers, and how love could be a teacher and a guide to what is right and good in life. The range of angles seemed limitless at times, and we enjoyed simply playing with ideas, some that led only to silence, but most led to even more questions.

From these discussions, it became clear that for Sir John Templeton, love was not a fleeting warm feeling, but a serious, beautiful, and majestic force. It was a key part of the fabric of reality and of each individual's path toward knowing God. Several times, Sir John would provocatively and insightfully ask, "Perhaps unlimited love and God are the same thing?" For me, a question like this is simply so big it is hard to comprehend all the answers that could follow from it.

In our various times together, I could tell from both the frequency and the kinds of questions that were being asked that the search for new insights and understandings of love were very centrally important to Sir John in his daily life. He shared writings on the subject from a wide range of religious traditions and asked questions that were far deeper and wider than I was ever accustomed to considering. "How does the practice (and struggle?) of loving transform each one of us?" "Is gratitude yet another form of love?" and "Is it ever possible to give too much love away?" prompted thinking on where the limits of love may be, and if there may be real limits at all beyond those that are rendered real by human limits and imperfections. In many conversations, Sir John spoke explicitly of "unlimited love" as perhaps the ultimate reality, and in this area, as in so many others he explored during his life, he was a genuine pioneer. As far as I could tell, he was never content to stop searching. Each answer led to several more questions, and to Sir John, this represented progress and accumulating, slowly but steadily, more spiritual wealth. In the several years when I would visit Sir John, his determination to explore boundaries of love was steady, and growing more and more intense. Sometimes, he would broach sensitive or unimaginable topics—such as "Should we even love Hitler?"—and later conclude, "Absolutely, we should aim to love each person unconditionally, even if we don't love what he did." Not many people whom I know would have the courage to comprehend such love.

Sir John's interest in love was a lot more than academic or intellectual. He aimed to put his learning into practice and test it. This was especially evident in meetings when the subject of love was being discussed. You could sense in him what seemed like an ongoing struggle, to love purely, totally, and unconditionally. That meant a full focus of attention, careful listening, and entering into a fellowship of wondering, exploring minds. Toward the end of his life, one could see love beaming in his eyes. Still, he battled to express more love, and you could sense the ongoing struggle. No matter how deep, it was not yet deep enough. A sense of profound gratitude to others was one of the many fruits of his ongoing effort, but I think that Sir John felt that there were many other fruits that had yet to be discovered. If the essence of God is love, Sir John was in the process of catching glimpses of the divine, and coming into communion with this greatest of ultimate realities, even as he lived within the limits of earthly reality.

I am grateful for the time that Sir John spent with me and that he chose to spend much of it discussing love. The invitation to ponder the mystery of unlimited love transformed my life. I have learned that love not only binds people and things together but that it allows the individual to transcend ordinary limits. Curiously, it creates and compounds its own energy. Just as "with God all things are possible," I learned that with love, many, many things are possible, and perhaps much more than most people may first realize.

A Brief Epilogue:
Some Quotations For Reflection

IN THIS FINAL CHAPTER I have taken a set of quotes from Sir John's
writings having to do broadly with the concept of Unlimited Love as
Ultimate Reality, and contextualized them within his continuing urgent
concern that religions in their plurality understand better and abide
more consistently in their spiritual principle of Unlimited Love, rather
than engage in destruction. Sir John saw this as the absolutely essential meaning of "progress in religion" or "spiritual progress," and he
dwelled on it heavily in his early and most essential book, *The Humble
Approach: Scientists Discover God* (1981/1995), and then later in *Possibilities for Over One Hundredfold More Spiritual Information: The Humble
Approach in Theology and Science* (2000a), which echoes *The Humble
Approach* but transposed the content into the form of questions. These
are, in my opinion, his two most foundational works. This theme of
"progress in religion" as progress in Unlimited Love is central to *Agape
Love: A Tradition Found in Eight World Religions* (1999), *Pure Unlimited Love: An Eternal Creative Force and Blessing Taught by All Religions*
(2000b), and *Why Are We Created?* (2003). The latter four works were
all written within a period of just several years, from 1999 to 2003, and
together they can all be taken as powerful affirmations of Sir John's
primary concern for a human future free of religious warfare.

The Ground of Being/Ultimate Reality
"Paul Tillich, a Christian theologian, sometimes referred to God as
the ground of being. But Mary Baker Eddy, Charles Fillmore, and
Ernest Holmes went further to suggest that matter may be an outward

manifestation of divine thought, and that the creative spirit called God is the only reality." (1981/1995, 21)

"If God is infinite, then it follows that all other reality is dependent on Him and cannot exist apart from Him. Matter and energy may be only contingent manifestations of God. Space and time may be only manifestations of God. We should not think of matter and energy as created by God but as now utterly independent of God. That would mean that God is not 'all in all' the creative Ground and Sustainer of all that is. Matter and energy may be only creaturely manifestations of the universal Creator. While God does not need the universe to be God, the universe may need to be increasingly supported and enfolded in His presence and power to be what it is. Maybe it can only exist in and through God." (1981/1995, 21)

"The excitement and importance of scientific study of nature and the cosmos are enhanced (not reduced) if we conceive of each discovery as a new revelation of a reality deriving from and grounded in God. When these new discoveries point more to the nonexistence of matter, it becomes easier to think of matter and spirit as a unity." (1981/1995, 21)

"As even 'a little part of him,' we realize the mutual unity of God and his creation. We realize that our own divinity arises from something more profound than merely being 'God's children' or being 'made in his image.'" (1981/1995, 22)

"God is Creator of the universe of time and of men. Creation proceeds from idea to word to sensory data. The invisible Creator is the universal spirit, the causative idea, which sustains and dwells in all He created and is still creating. The orderliness and lawfulness of nature and of the spirit reveal God to man." (1981/1995, 24)

"Maybe the more we create, the more in some ways we are like God, especially if, like God, we create out of love." (1981/1995, 24)

"Emanuel Swedenborg wrote that nothing exists separate from God. If God is infinite, then nothing can be separate from Him. In Him we live and move and have our being. God, he claimed, is all of you and you are a little part of God." (1981/1995, 17)

"Twenty-five centuries ago Xenophanes and twelve centuries ago

Shankara taught that nothing exists independently of God and that God is immeasurably greater than all time and space, let alone the visible earth." (1981/1995, 3)

"Every person's concept of God is too small. Through humility we can begin to get into true perspective the infinity of God. This is the humble approach." (1981/1995, 3)

"Is god ultimately the only reality—all else being fleeting shadow and imagination from our very limited five senses acting on our tiny brains?" (2000a, 20)

"Some astronomers at work discovering the vast complexities of the macrocosm, and nuclear physicists investigating the awesome variety of the microcosm, are concluding that the universe bears the hallmarks of intelligent design." (2000a, 66)

"As St. Paul remarked to the Athenians, quoting a Hellenistic poet (Acts 17:28), 'In him we live and move and have our being.'" (2000a, 81)

"So, if there is a phenomenal universal force, for example, in gravity, in the light spectrum, can there not also be a tremendous unknown, or non-researched, potency, or force, in unlimited love?" (2000b, 35–36).

"Could unlimited love be described as a creative, sustaining energy? When we embrace our creative energy, can we draw, from the universal Source, a tremendous spiritual energy matrix into many areas of our lives? Does a divine fountainhead of love exist in the universe in which degrees of human participation are possible?" (2003, 87)

"Those who are philosophically inclined may find it helpful to understand God's unlimited love as the original and ongoing basic creative force of the universe. This love was present before the beginning, and it continues to hold all things together. Our fleeting human emotions and perceptions are in fact mere glimpses of God's perfect love." (2000b, 19)

The Ontological Generality
(or The True Self vis-à-vis The False Self)

"God is the source of love. Love cannot flow unless it also flows out. The Spirit of God is like a stream of water, and His disciples are like many beautiful fountains fed by this river of waters. Each of us is such a fountain, and it is our task to keep the channel open so that God's Spirit can flow through us and others can see His glory. Without God we are not likely to bring forth any good. If we think too much of the visible world or trust in our own ability, we become like a clogged fountain. We will never learn to radiate love as long as we love ourselves, for if we are characterized by self-concern, we will radiate self-concern." (1981/1995, 110)

"We should seek always to let God's love shine forth like the light inside an electric bulb illuminating all our habitation." (1981/1995, 110)

> When Jesus was asked what is the greatest law, he said: Thou shalt love the Lord thy God with all thy heart, and with all thy soul, and with all thy mind. This is the first and great commandment. And the second is like unto it, Thou shalt love thy neighbor as thyself. On these two commandments hang all the law and the prophets. (Matthew 22:37–40)

"This can be researched as a basic law of the spirit. A person who applies this law often finds her life revolutionized. Opening our hearts allows god's love to flow through us like a mighty river. If we love as god loves us, we learn to love every person without exception. The happiest people on earth seem to be those who give love wholeheartedly always." (2000, 160–61)

Secular Humanism

"Those choosing the humble approach shun the colossal conceit of the atheist, who believes that the material world is the only reality and that it is totally explicable without a creator." (1981/1995, 22)

"Marx believed that his kitchen table was composed of solid matter. Now even scientists living in formerly Marxist nations believe a table is, to a large degree, space or nonmatter. Its solid appearance comes only from responses by man's senses to billions of invisible electrons and

wavelike particles vibrating at the speed of light in various patterns."
(1981/1995, 23)

"The humble approach rejects all self-centered philosophies, especially those brands of humanism that teach that man is the end purpose
of creation. In humility let us admit that God's awesome creative process is likely to continue even if humans should disappear from the face
of the earth. Humanism is egotistical because it encourages men to think
that mankind is itself the ultimate concern. Rather than worshipping the
universal Spirit creating the universe, humanists worship the creatures.
Communists go even further to worship governments which are created
by creatures. Today they wonder why the children of Israel in the wilder-
ness worshipped a golden calf which they themselves had created. Is it
any wiser to worship a government created by men?" (1981/1995, 114)

"An atheist who is sure there is no God is really a pitiful person
because he is too egotistical to admit his limitations and insignificance."
(1981/1995, 45)

The Threat of Religious Arrogance

"Some people have said that religion causes wars, that the diversity of
beliefs will always be divisive. But the fact is that such wars were brought
about for a number of reasons, not the least of which was egoistical and
unwarranted claims to proprietary rights over the knowledge of God.
History reveals that the great havoc and suffering were caused not by
religion, but by men who thought that their concept of God was the only
one worthy of belief." (1981/1995, 45)

"No one should say that God can be reached by only one path."
(1981/1995, 46)

"The human ego has been the curse of religious denominations
for thousands of years. In every major religion wars have been fought
about differences in creed. Nations or tribes have exterminated others
because they worshipped different gods or the same god as taught by
different prophets. This is human ego run wild. Let us humbly admit
how very small is the measure of men's minds. This realization helps

to prevent religious conflicts, and obviates attacks by atheists against religion. Moreover, humility of this kind opens more minds to the idea that science supports and illuminates religion." (1981/1995, 47)

"Through humility we can avoid the sins of pride and intolerance and avoid especially harmful religious strife because it is unlikely any religion could know more than a tiny bit about an infinite god." (2000a, 8)

"To seek to persuade all people to believe in one perspective would be a great tragedy. The wide diversity of faiths and theologies is a precious aspect of the richness of religion despite the fact that from a scientific point of view it can seem flawed and relativized. What is needed may be creative interaction and competition in a spirit of mutuality, respect, and shared exploration. Were a drab uniformity to be accepted, the possibility of spiritual progress would be diminished. Learning to live with and learn from a rich multiplicity of spiritual perspectives is a step forward, because the more we know, the more we know we do not know. This helps give life creativity and joy." (2000a, 18)

"It can be a religious virtue reverently to cherish scriptural beliefs and to study them with utmost seriousness. But of course a reverse side of this virtue can be the vice of intolerance. It is easy to become intolerant if we are not diligent to guard our minds actively to be humble and to remember that despite differences in religious traditions we all have profoundly limited concepts with respect to the vast divine realities. Can love and the vastness of divinity reduce our differences as we seek to understand by a variety of different ways and through many various traditions? Can diligence in humility help heal conflict between many communities holding different religious points of view?" (2000a, 28)

"History reveals that great havoc and suffering were caused not by religion but by people who thought their own concept of god was the only one worthy of belief." (2000a, 29)

"To best deal with conflicts over dogmatism, possibly we can benefit most by listening carefully and respectfully rather than arguing." (2000a, 36)

"Should we be gentle, kind, and sympathetic toward new prophets even though they bring new ideas strange to us?" (2000a, 38)

"All of the world's great religions, to varying degrees, both teach and assume the priority of love in religious practice. To put it another way, whether consciously or subconsciously, the world seems to have determined that any system of beliefs that teaches or tolerates hatred or even apathy toward others does not deserve to be considered a religion in the first place." (1999, 2)

"The rich variety of world religions creates a tapestry of amazing beauty—a testimony to the essential spiritual nature of our human existence. And yet, within this amazing, and sometimes fascinating diversity can be found an equally amazing unity, the basis of which is 'love.'" (1999, 2)

The Eternal Soul

"However, for the sake of simplicity, we will consider *God* to be the infinite creator of the cosmos. If He is truly infinite, then nothing exists apart from Him, and all other realities are created reflections of Him. *Soul* signifies that divine infusion which is unique to each human being. Most major religions teach that each soul is immortal and can be educated. *Mind* is defined as the strategic link between soul and body. Mind is complex and miraculous but temporary. The human *body* is defined as a temporary physical dwelling for the mind and soul. In the light of scientific analysis all bodies, human and subhuman, are only evanescent arrangements of forces and wave patterns.

"Obviously the human body, especially the brain, has great effect on the mind. When we are tired or sick, for instance, we often think less clearly. But basically the determining influence flows in the other direction, from the unseen to the seen, from God to soul, to mind, to body. These four realities, but probably the lesser three derive from the first." (1981/1995, 103–4)

"We should never forget that the body is only a temporary configuration of wave patterns." (1981/1995, 104)

"The great question is this: How much progress can our soul make before our body becomes uninhabitable? To progress is to increase our love of God, our understanding of God, and our love for his children.

Our body has a physical reality, but it is only a temporary shell. Death destroys only that which is fit for destruction. The butterfly developing in the chrysalis in due time splits and abandons the dead chrysalis and flies away on wings of amazing beauty undreamed of by the chrysalis or the caterpillar." (1981/1995, 91)

Sir John quotes often from Hindu sources, which fascinated him almost as much as Christian. Here is a passage he cites from *The Bhagavad-Gita* (1981/1995, 91): "I am the Atman that dwells in the heart of every mortal creature."

"Is our body like our house? Also, is our mind? Let us take care of it while we live in it, using our mind to help develop our soul." (2000a, 143)

Life on Earth and Its Sufferings in the Service of Love Growth
"How could a soul understand divine joy or be thankful for heaven if it had not previously experienced earth? How could a soul comprehend the joy of surrender to God's will, if it had never witnessed the hell men make on earth by trying to rely on self-will or to rely on another frail human being or on a soulless man-made government?" (1981/1995, 93)

"Maybe the earth was designed as a place of hardship because it is the best way to build a soul—the best way to teach spiritual joy versus the bodily ills. Why was it said that into every life some rain must fall? It is apparent that sometimes a great soul does not develop until that person has gone through some great tragedy. Let us humbly admit that God knows best how to build a soul." (1981/1995, 93)

"As a furnace purifies gold, so may life purify souls." (1981/1995, 94)

"Of course all of us should work for self-improvement by prayer, worship, study, and meditation. But one of the laws of the spirit seems to be that self-improvement comes mainly from trying to help others— especially from trying to help others enjoy spiritual growth. Growth comes by humbly seeking to be a more useful tool in God's hands." (1981/1995, 95)

"One of the major lessons to learn while on earth is that building our heaven is up to us. Emanuel Swedenborg wrote that we will not be in heaven until heaven is in us. Here on earth we can begin to receive the life and spirit of heaven within us." (1981/1995, 96)

"It cannot be repeated often enough that we can learn to radiate love; but first we must practice using loving words and loving thoughts. If we can keep our minds filled with good thoughts of love, giving, and thanksgiving, they will spill over into our words and deeds." (1981/1995, 107)

"The Christian religion speaks often about the way in which the human experience on earth is to function as preparation. To progress spiritually can be to increase our love of god, our understanding of god, and love for his children." (2000a, 132)

"Why should we waste even one day? At the end of the day, can we say we have learned to radiate to all pure unlimited love and helped our neighbor to learn the joy of giving love?" (2000a, 136)

"These sayings could well be called laws of love. If we radiate love, we receive joy, prosperity, happiness, peace, and long life in return." (2000a, 138)

Sir John's Great Purpose and Why Unlimited Love as Ultimate Reality Was His Path

Sir John believed in progress, but he was uncertain of the future of the human species, and even believed it a mark of arrogance to think that humans are somehow at the center of the universe. He was a realist about frail, unreliable, proud, and arrogant human nature.

Thus, he understood that without a true renaissance of understanding and abiding in Unlimited Love, nothing is secure about the human future in a time of technologies of mass destruction. His hope was to revivify ancient traditions of religion and spirituality with the ideal of Unlimited Love, in part by pointing out that, in fact, this Unlimited Love is plausibly Ultimate Reality, the very essential sustaining Ground of Being that underlies all that exists.

I believe that Sir John the Tennessee mystic knew this Unlimited Love as Ultimate Reality to be in fact subjectively valid. But he did not see much hope of transforming modern cultures in this constructive direction without the powerful findings of physics. I have argued on Sir John's behalf that this idea of Unlimited Love as Ultimate Reality is plausible. Sir John did not necessarily believe that we would get quite to the bottom of Ultimate Reality for a long time. But he would certainly have wished to assert that the idea of Unlimited Love as Ultimate Reality is a good and plausible one—even as good or better than the alternatives.

In the meanwhile, he emphasized that people have tremendous spiritual transformations in the direction of more extensive and effective love when they report experiences of God's love. He stressed that human flourishing is optimal in the discovery of the true self in the ontological generality of love of God and neighbor as self. In other words, he was all for the love of self, but he wanted us to love ourselves rightly and truly in the context of the generality. And Sir John was hopeful that science would eventually touch on Ultimate Reality and Unlimited Love as the great sustaining matrix of divine Mind that is the substrate of everything that is.

I have tried in this volume to organize loyally these core arguments that Sir John felt to be of such importance. If those who read this book gain some greater insight into Sir John's core big idea, then I have succeeded. Throughout I have labored to draw directly on Sir John's writings, and here and there to extrapolate from them and extend them. But throughout I have been guided by one principle: *What would Sir John say?* Others will come along in the future and improve upon this book, as I would hope. May this work, *Is Ultimate Reality Unlimited Love?* be worthy of a true visionary mind whose prophetic thought world was far ahead of its time. I again want to thank my honorable friend John M. Templeton Jr., MD, for allowing me the honor of trying my hand at this endeavor.

Appendix

John M. Templeton

Box N.7776, Lyford Cay, Nassau, Bahamas

Telephone: (242) 363-4904 · Fax: (242) 363-4880

August 3, 2001

VIA FACSIMILE   **(216) 368-8713**

Dr. Stephen G. Post
Prof. Of Biomedical Ethics,
        Religious Studies, and Philosophy
Case Western Reserve University
School of Medicine
10900 Euclid Avenue
Cleveland, OH 44109-4976
USA

Dear Stephen:

Many, many thanks for your long and helpful memorandum you sent to me and Mrs. Marchand July 30 about the proposed *"Institute for Unlimited Love Research"*.

The somewhat related subject of altruism is important also, but if Fetzer approves, altruism could just be included as one of the many consequences of unlimited love.

I am please indeed, by your extensive plans for research on human love. I will be especially pleased if you find ways to devote a major part, perhaps as much as one-third of the grant from the Templeton Foundations, toward research evidences for love over a million times larger than human love. To clarify why I expect vast benefits for research in love, which does not originate entirely with humans, I will airmail to you in the next few days some quotations from articles I have written on that subject.

Dr. Stephen G. Post
August 3, 2001
Page 2.

Is it pitifully self-centered to assume, if unconsciously, that all love originates with humans who are one temporary species on a single planet? Are humans created by love rather than humans creating love? Are humans yet able to perceive only a small fraction of unlimited love, and thereby serve as agents for the growth of unlimited love? As you have quoted in your memorandum, it is stated in John I that *"God is love and he who dwells in love dwells in God and God in him"*.

For example, humans produce a very mysterious force called gravity but the amount produced by humans is infinitesimal compared to gravity from all sources. Can evidences be found that the force of love is vastly larger than humanity?

Can methods or instruments be invented to help humans perceive larger love, somewhat as invention of new forms of telescopes helps human perceptions of the cosmos? What caused atoms to form molecules? What caused molecules to form cells temporarily? What caused millions of cells to combine into productive plants? Can humans research non-human love by studying why hundreds of worker bees devote their lives to feeding their queen and their drones? Could love be older than the Big Bang? After the Big Bang, was gravity the only force to produce the galaxies and the complexity of life on planets?

Do you remember the old song, *"Love is the answer. Love is the answer to it all"*? Does love help to answer the basic questions of why, why, why? For example, why is there something rather than nothing? Why are we here temporarily on this little planet?

Turning now to evidences of the benefits of love in human life, there are many evidences that anger is never solved by anger but is solved by love. If bigger boys tease a smaller boy in order to make him angry, is the problem solved quite quickly if the little boy learns to reply quietly to every taunt by simply saying 'I love you, I love you'?

Dr. Stephen G. Post
August 3, 2001
Page 3.

In your memorandum you invite suggestions for leaders who might be helpful to the Unlimited Love Research Institute. Therefore, I am mailing to you today a copy of the Forbes Magazine list of all the world's billionaires on the possibility that you or your friends or some of the Templeton Foundation Board of Advisors could help you explain the importance of your institute to potential donors and potential trustees.

To create even wider enthusiasm for your institute, I hope you will submit each quarter to Radnor and to me and to our newspaper about 300 words about progress in your important program. If your progress is reported quarterly, that may automatically influence as many as 20,000 readers, already having some interest in the cooperation of science and religion.

God bless you,

John M. Templeton

JMT:mes

cc: Dr. Charles L. Harper, Jr.
      Mrs. Judith Marchand
      Dr. John M. Templeton, Jr.

CASE WESTERN RESERVE UNIVERSITY

August 30, 2001

Sir John M. Templeton
Box N-7776, Lyford Cay
Nassau, Bahamas
Fax: (242) 362-4880

answering for Sir John SGB

Dear Sir John:

This letter is written simply to express my appreciation and gratitude to you for your many beneficial comments and suggestions in the process of establishing the Institute for Research on Unlimited Love.

Please know that I will fully implement your wishes, and that I look forward to our meeting at 9:30 a.m. on Monday, October 1, in Nassau. It is my intention to attentively gain from your insight and vision so that the Institute is absolutely on target.

Toward this end, I want to mention to you that on Monday, October 1, many of the designated Research Area Consultants for the Institute, along with myself, will be convening an initial planning session in the Harbor Island Room at the Nassau Marriot, commencing with a noon luncheon. We will probably work throughout the afternoon and into the evening.

It would mean a great deal if you could join us anytime in the afternoon, or perhaps for an evening session over dinner. I know that representatives from the Templeton Foundation will also be present, including Judy Marchand. All are of course more than welcome.

I believe that as you get to know me more, you may discover that we are very much thinking alike, and that I am always amenable to suggestions and correction. I view my role as one of a facilitator for a project that emerges from your ideals and creativity, and I therefore earnestly intend to follow your vision to the fullest.

Incidentally, I attach yet another effort to arrive at an increasingly full definition of your concept of *unlimited love*. Like prior efforts, it needs to be enhanced. You are of course, as the creator of this terminology, welcome to edit as needed. I will happily implement!

Thank you very much.

With Every Best Wish,

Stephen G. Post, Ph.D.
Institute for Research on Unlimited Love

Cc: Dr. Jack Templeton, Judith Marchand, Dr. Chuck Harper

Dear Stephen, Congratulations on an excellent beginning! Hopefully you can craft a "definition" with details implying that "unlimited love" is only from humans toward humans. Possibly love toward all or on more universe than growth. Possibly humanity is just the latest creation on one planet of the unlimited creative reality called LOVE?

Center for Biomedical Ethics

Case Western Reserve University   Phone 216-368-6196   MetroHealth Medical Center Phone 216-778-8497
School of Medicine   Fax   216-368-8713   Rammelkamp Center
10900 Euclid Avenue   Email xx245@po.cwru.edu   2500 MetroHealth Drive
Cleveland, Ohio 44106-4976   Cleveland, OH 44109-1998

You may want to read carefully my little book "Pure Unlimited Love". God bless you. John M. Templeton, 31-8-01

August 30, 2001 (draft 4)

THE INSTITITUE FOR RESEARCH ON UNLIMITED LOVE

The Institute for Research on Unlimited Love has been established as an independent non-profit entity located at the School of Medicine of Case Western Reserve University. The Institute takes the language of *"unlimited love"* from the thought and writings of Sir John Templeton, and draws its major initial support from the John Templeton Foundation. The Institute exists to (a) further our knowledge of unlimited love at the research interface of science and religion, (b) provide educational opportunities, and (c) disseminate findings in such a way as to positively influence the world.

The Institute offers the following preliminary definition of unlimited love:

Unlimited love is, within spiritual and religious traditions, generally considered a divine attribute and an energy integral to all reality, one in which creaturely participation is possible as the fullest experience of spirituality, giving rise not merely to sentiment but to active works of love.

Unlimited love at its core is an enduring affirmation of the value and even sacredness of the existence of all others; it is the full recognition that the being of all others means as much or more to me than my own being. Depending on the circumstances of the other, unlimited love is variously expressed through empathic understanding, compassion, care and altruism, dialogue, empowerment, celebration, communion, forgiveness, and other manifestations. (In much of ordinary human experience, these expressions of love typically do not reach the heights of unlimited love.) At core, unlimited love is an affirmation that acknowledges for all others the absolutely full significance that, because of egoism, we otherwise acknowledge only for ourselves. This migration at the

1

center of our being to all humanity is the core element in all forms of unlimited love. While unlimited love generally welcomes a reciprocal response from others, it is not dependent on such mutuality. Unlimited love as a positive energy of course assumes all restraints on harm to humanity.

Just as human beings endeavor to understand and harness the power of the wind, the atom, and gravity, they might also endeavor to understand and facilitate the energies of unlimited love. To do so requires rigorous research in all disciplines, coupled with a willingness to dialogue across disciplinary boundaries.

An immediate goal of the Institute is to disseminate a Request for Research Proposals by February 2002 for scientific research projects of the highest methodological quality as assessed by rigorous peer review. These projects, which will represent a variety of disciplines and fields, must engage seriously with the ongoing dialogue between science and spirituality-religion.

John M. Templeton
Box N7776, Lyford Cay, Nassau, Bahamas
Telephone: (242) 362-4904 · Fax: (242) 362-4880

RESEARCH OPPORTUNITIES TOWARD BENEFICIAL DISCOVERIES ABOUT UNLIMITED LOVE

TO: Stephen Post from John Templeton September 1, 2001

A complete definition of "unlimited love" is not possible by humans, because perceptive abilities of humans are so limited. Instead we can research possible pathways toward unlimited love.

Unlimited love may be billions of times more vast than any one temporary species on a single planet can yet comprehend. Already some evidences are perceptible by humans pointing toward unlimited love as a power aggregating particles into atoms and molecules and cells and plants and animals and human brains and families and nations. At each higher level of complexity the whole is greater than the sum of parts. Creativity accelerates and variety accelerates.

Gravity is easier to measure than love; but the power of love may be more creative, more timeless, more vast, more beneficial.

As we develop methods to increase our perceptions, we may discover more about unlimited love and the biblical words "God is love; and he who dwells in love, dwells in God and God in him".

As humans invent new perceptive abilities, we may research other evidences of unlimited love. Evidences of love radiating from people, provide data for experimental science research, as a fruitful beginning. As just one example, research is needed on why it is impossible to give away too much love. The more you try to give love, the more love you have left to give.

September 4, 2001

To: Judy Marchand
610—687-8961

Sir John Templeton
(242) 362-4880

Dear Sir John:

I too spent some time this weekend rethinking working definitions of *unlimited love*. I did reread your book on the subject.   I attach a revision.

Now I am reading your fax of this morning (Tuesday Sept. 4), and will try to incorporate it in fitting manner.

Thanks very much.

Sincerely,

Stephen Post

P.S. I look forward to our discussions on this in Nassau both personally, and at the advisory meeting.

Cc. Chuck Harper

Institute for Research on Unlimited Love
Located at the School of Medicine, Room T-401
Case Western Reserve University
10900 Euclid Avenue
Cleveland, Ohio 44106-4976
(216) 368-6205

Stephen G. Post, Ph.D.
President

September 7, 2001

VIA CERTIFIED MAIL
RETURN RECEIPT REQUESTED

Internal Revenue Service
Cincinnati, OH 44099
STOP 422 UNIT 34

RE: **Reporting Change of Corporate Name**
Former Name: Institute for Unlimited Love and Altruism Research
New Name: Institute for Research on Unlimited Love
Employer Identification Number: 34-1961143

Dear Sir or Madam:

Please be advised that Institute for Unlimited Love and Altruism Research, an Ohio nonprofit corporation, has changed its corporate name to Institute for Research on Unlimited Love effective August 9, 2001. We therefore respectfully request that you change your records to reflect this change of corporate name.

Additionally, to ensure your receipt of this notification, please file-stamp the enclosed copy of this letter and return it to me in the envelope provided.

Very truly yours,

Carol A. Adrine
Assistant Secretary

cc: Paul H. Feinberg, Esq.
Jonathan J. Hunt, Esq.

Department of the Treasury
Internal Revenue Service

CINCINNATI, OH 45999

In reply refer to: 1765826676
Nov. 02, 2001 LTR 252C
34-1961143 000000 00
01370

INSTITUTE FOR RESEARCH ON UNLIMITED
% STEPHEN G POST
10900 EUCLID AVE CWRU MED SCHL T401
CLEVELAND OH 44106-1712004

Taxpayer Identification Number: 34-1961143

Dear Taxpayer:

Thank you for the inquiry dated Sep. 07, 2001.

We have changed the name on your account as requested. The number
shown above is valid for use on all tax documents.

If you have any questions, please call Ms. Gloria Long at
859-669-4131 between the hours of 6:30AM and
2:00 PM EST. If the number is outside your local calling area,
there will be a long-distance charge to you.

If you prefer, you may write to us at the address shown at the top
of the first page of this letter.

Whenever you write, please include this letter and, in the spaces
below, give us your telephone number with the hours we can reach you.
Also, you may want to keep a copy of this letter for your records.

Telephone Number (   )_____ Hours_____

Thank you for your cooperation.

Sincerely yours,

Alan Berger

Alan Berger
Chief, Research and Perfection Br.

Enclosure(s):
Copy of this letter
Envelope

John M. Templeton
Box N-7776, Lyford Cay, Nassau, Bahamas
Telephone: (242) 362-4904 · Fax: (242) 362-4880

August 5, 2003

VIA FACSIMILE   (216) 368-8713

Dr. Stephen G. Post
The Institute For Research On Unlimited Love
Prof. Of Biomedical Ethics,
    Religious Studies, and Philosophy
Case Western Reserve University
School of Medicine, Room 214
10900 Euclid Avenue
Cleveland, OH 44106-4976
USA

Dear Stephen:

Many thanks for your excellent 10-page report dated July 21 about your multitude of activities in the *Institute for Research on Unlimited Love*.

Especially, I am impressed with your success in National Public Radio. Millions of intelligent people in more and more nations enjoy listening to these intellectual radio programs, who want to make better use of the time they devote to commuting or bathing or other essential activities, when the only way to learn is by listening. Your friends who are already experts in public radio can help grantees of yours to develop programs to help those executives who are always searching for more programs. Especially, they search for programs which include emotional stories such as *unlimited love*. The time may come when the program managers in many nations are willing to pay your experts to put on tape stories about *unlimited love* which have wide emotional appeal or scientific discovery.

Dr. Stephen G. Post
August 5, 2003
Page 2.

Congratulations on getting so many busy people to attend big
conferences at their own expense. However, I wonder if, for a small
fraction of that expense, you could reach an even larger audience by
inviting our newspaper to mail to their readers, about twice a year, a
supplement of 10 or 20 pages about the exciting activities of your *Institute
for Research on Unlimited Love*.

God bless you,

John M. Templeton

JMT:mes

cc: Dr. John M. Templeton, Jr. (via facsimile 610 687-4676)
      Dr. Harold G. Koenig (via facsimile 919 383-6962)
      Dr. Karl W. Giberson (via facsimile 617 745-3905)

John M. Templeton
Box N-7776, Lyford Cay, Nassau, Bahamas
Telephone: (242) 362-4904 · Fax: (242) 362-4880

October 28, 2003

VIA FACSIMILE   (216) 368-8713

Dr. Stephen G. Post
Case Western Reserve University
The Institute For Research On Unlimited Love
School of Medicine, Room 214
10900 Euclid Avenue
Cleveland, OH 44106-4976

Dear Stephen:

Many thanks for the helpful information in your October 17 letter. Congratulations on the excellent progress you reported at the Foundation meetings in Cambridge.

I will be especially interested if you or your colleagues discover any articles on the possibility that all forms of life may be manifestations of unlimited love instead of the usual concept that love should be studied as one of the many aspects of many humans. In other words, among vast numbers of other scientific studies, I would welcome copies of any studies about why after the Big Bang the cosmos did not become merely an inanimate cloud.

Your friend,

John M. Templeton

JMT:mes

cc: Dr. John M. Templeton, Jr. (via facsimile 610 687 4676)

John M. Templeton
Box N-7776, Lyford Cay, Nassau, Bahamas
Telephone: (242) 362-4904 · Fax: (242) 362-4880

October 30, 2003

VIA FACSIMILE   (216) 368-8713

Dr. Stephen G. Post
Case Western Reserve University
School of Medicine, Room 214
10900 Euclid Avenue
Cleveland, OH 44106-4976
USA

Dear Stephen:

As usual, I am searching for ideas about how your beneficial work can attract more enthusiasm and more donors.

Do your Advisors think it will be helpful for you to subsidize a weekly email letter for thousands of possible donors, each letter being on a different aspect of *unlimited love*, beyond the love which is a product of humanity?

Probably, top quality articles can be found about how *unlimited love* may have been the creative force 15 billion years before humanity began, and how *unlimited love* may be the force behind (rather than the product of) the amazing acceleration of creativity.

Also, if you find fascinating news on this subject, it could be used in our newspaper called "*Research News & Opportunities*" and in other intellectual journals.

God bless you,

John M. Templeton

JMT:mes

cc: Dr. John M. Templeton, Jr. (via facsimile 610 687-4676)

John M. Templeton
Box N7776, Lyford Cay, Nassau, Bahamas
Telephone: (242) 362-4904 · Fax: (242) 362-4880

FROM JOHN M. TEMPLETON to Dr. Stephen G. Post
November 10, 2003

Astronomers have helped us to see how small minded were ancients
who spoke of divinity as if it resembled a king above the sky. Are
humans today equally small minded if they write about love as if love is
only a feature of a single recent race on a single planet?

What science methods can lead to help discoveries about love and its
benefits? Why do quarks form patterns as atoms and atoms form
molecules? Can this be called the science of unlimited love?

JMT:mes

John M. Templeton
Box N-7776, Lyford Cay, Nassau, Bahamas
Telephone: (242) 362-4904 · Fax: (242) 362-4880

February 25th, 2004

Dr. Stephen Post
Case Western Reserve University
School of Medicine, Room 214
10900 Euclid Avenue
Cleveland, Ohio 44106-4976

Dear Stephen:

Congratulations on your diligent and extensive work for your foundation to encourage scientific methods for discovering over one hundredfold more of aspects and benefits of unlimited love. Your progress report which my son discussed at our meeting yesterday seems to imply that your program is limited to the type of love called Agape. My intention when making a major grant to your program was that it would encourage the top one millioneth intellectually of the world's people to become enthusiastic about science research on every form of love, not only all types of human love but also increasing evidences that love may be the driving force in all creativity even before the Big Bang.

For example, evidence is increasing that creativity itself is a form of love and so is magnetism and gravity. The New Testament says that "for God is love and he dwells in love dwells in God and God in him". I hope that your program can be expanded to include a new encyclopedia about all science methods which have ever been applied to discover more about the benefits and aspects and varieties of love. Love has been a central theme of all major religions for thousands of years, but until the latest two centuries methods of science have not been applied for human discoveries of over one hundred fold more about love.

God Bless you and your important program.

John M. Templeton

05-27-04   16:02   FROM-first trust bank ltd   +242 362 4814   T-757 P.001/001 F-660

THE INSTITUTE FOR RESEARCH ON
UNLIMITED LOVE

ALTRUISM. COMPASSION. SERVICE

The Good Samaritan is a universal symbol
of love for all humanity.

May 27, 2004

2 42 - 362 - 4880

Dear Sir John:

I hope you are well. I have thought a lot about your note of
10 May regarding a quotation for the Institute. I mention herein
two such quotations, which I may use in tandem.

The first is:
"In the giving of self lies the unsought discovery of self."

The second is:
"The best way to predict the future is to invent it."

Together, these work synergistically. I like them side by side.

Interestingly, the second one is now found now on a website of so-called
"famous" quotations. This is not quite believable, since both emerge
from my own various media and column writing, and are, then,
Institute creations! Of course I do not want to take credit for them,
but both have really caught on and say a lot to people. I am able
to be very effective and impassioned with them.

This is good timing for your thoughts, since I am spending the next
couple of days revising the PR packet for the Institute.

Thanks. I am well and look forward to your response.

All best wishes

Stephen

Cc: Fax Dr. John M. Templeton, Jr., M.D. 610-687-8961

*Yes, I approve both of these excellent
quotations. John Templeton
27-5-04*

John M. Templeton
Box N-7776, Lyford Cay, Nassau, Bahamas
Telephone: (242) 362-4904 · Fax: (242) 362-4880

May 2, 2005

VIA FACSIMILE (216) 368-8713

Dr. Stephen G. Post
Case Western Reserve University
The Institute For Research On Unlimited Love
School of Medicine, Room 214
10900 Euclid Avenue
Cleveland, OH 44106-4976
USA

Dear Stephen:

Your wisdom will be welcomed by me and your other colleagues on the following five ideas: -

1. Often people seek love to become happy. But really, happiness comes from giving unlimited love. The more love you can give the more you have left to give.

2. Unlimited love giving is over ten times more beneficial if focused more on ways to prevent rather than ways to alleviate suffering.

3. Giving materials often causes the receiver to remain childish, whereas opportunities to produce and to become self-reliant help the receiving to grow spiritually.

Dr. Stephen G. Post
May 2, 2005
Page 2.

4.   When a young child is able to listen, should a mother ask each morning, "How do you plan to use this new day to give more unlimited love and to whom and how?"

5.   Life is always a vast mystery, but enlightenment can come from enthusiasm to give unlimited love.

God bless you,

John M. Templeton

JMT:mes

cc: Dr. John M. Templeton, Jr. (via facsimile 610 825-2962)
      Dr. Charles L. Harper, Jr. (via facsimile 610 825-1730)
      Mrs. Judith Marchand (via facsimile 610 825-1730)
      Mrs. Pamela Thompson (via facsimile 610 825-1730)
      Dr. Gail Zimmerman (via facsimile 307 266-3846)

May 9, 2005

Sir John M. Templeton
Box N-7776 Lyford Cay
Nassau, Bahamas
Fax- (242) 362-4880

Dear Sir John:

Recently, on May 2, I received an important fax from you regarding
five ideas. May I take this opportunity to respond in some depth?

The first idea is highly significant: "1. Often people seek love to
become happy. Yet happiness comes from giving unlimited love. The
more love you can give the more you have left to give." In response to
this idea, I have drafted a white paper for you over the last week. It is
entitled "Benevolent Unlimited Love, Happiness, and Health: Rx 'Do
Unto Others.'" Were this to become the basis of a new request for
research proposals, I personally believe that we could change the world.
Please know that I have organized the 35 leading scholars around this
same idea, and worked with them since a year ago to produce a book
in press with Oxford University Press. In addition, the energy and
"flow" of love as a vital emotional reality underlying happiness and health
is the topic of my book, in preparation, with Random House. I have been
happily slaving away at this since May of 2004. It is, shall we say, my
core purpose in life.

The second idea is also highly significant: "2. Unlimited love giving is
ten times more beneficial if focused on ways to prevent rather than ways
to alleviate suffering." Indeed, this is the most wise and effective
distribution of such love. With this in mind, I have over the last six
months placed considerable focus on youth. How can we love our
children in such a way as to avoid indulgence, teach responsibility,
and mold them in ways that will make them the agents of unlimited love
in the future. The Institute has funded four new pilot projects on how
to implement love in the lives of youth so as to prevent their suffering
and allow them to realize the motto of the Institute, which is, as you
endorsed it, "In the giving of self lies the unsought discovery of self."
The projects sponsored are at the following sites:
*Harvard University School of Public Health
*The University of Pennsylvania

*The Search Institute
*The Presbyterian Church U.S./ Dr. Carolyn Schwartz

The third idea is vital as well: "3. Giving materials often causes the receiver to remain childish, whereas opportunities to produce and become self-reliant help the receiving to grow spiritually." Very much so. This is why my dad had me working in the ship yards along the south shore of Long Island from age 12, summer and weekends. By age 14 I was able to actually build small wooden sailboats. My pay was never more than $2.50 an hour. But in those ship yards, especially one in my home town of Babylon owned by a real tough fellow named Dave Suthard, I learned a lot about hard work and creativity. I was also given a small clamming boat with an outboard engine, and in high school days could make as much as $40 a day selling my clams by the bushel at the Bayshore docks. I attribute much of my successes in life, including here at Case, to this work ethic. I think there could be a serious research program in the topic area of "Giving Love Wisely and Efficiently so as to Encourage Spiritual Growth in Recipients." Sir John, our kids are overindulged and pampered to death, and many simply cannot actually *do* anything!

The fourth idea is laudable: "4. When a young child is able to listen, should a mother ask each morning, 'How do you plan to use this new day to give more unlimited love and to whom and how?'" Yes, and in fact, several months ago, after convening a dozen experts on "how to raise a child of love," this in collabortion with the Fetzer Institute, we were able to fund four projects on the topic of how to accomplish this. Moreover, I have just closed a matching arrangement with the Emory University Center for Religion on "the right love of a child."

The fifth idea is a great one: "5. Life is always a vast mystery, but enlightenment can come from enthusiasm to give unlimited love." This is very much the topic of my 120-page research area topic paper for the new "Religion, Spirituality & Human Flourishing" RFP that the Templeton Foundation has initiated. We have to be a little ambivalent about religion because it can bring out the very best in people and the very worst. Even the word "spirituality" is so overused as to become cliché. So, *enlightenment* is probably the best paradigm. Can we say that the more a person discovers the happiness and flourishing associated with giving unselfishly to others, the more he or she is likely to have some humble sense of the ultimate reality underlying all of existence? Sir John, all true and legitimate growth in knowledge often flows indirectly from a life of deep other-regarding love. In this sense, knowledge and wisdom are like happiness – they are all implicit in the love of others. I personally strive each day to pay special attention to just everyday folks, and to try to have a flow of consistency in how I treat people from the moment I rise until I go to bed (early, usually). It is this consistency, which we never fully achieve, that animates a certain mysterious emotional engagement in the energy of love, and this is what allows insight and creativity at their best. I do go to church most Sundays, but mainly see this as just another way of placing the priority during the week on love. To interrupt the flow of love energy in one's life is a serious matter. This does not mean that one fails to confront destructive behaviors, but one does so out of love rather than bitterness. I do believe that God is love, literally, and that to the extent we can even begin to be consistently loving toward everyone, without being limited by snobbery or an inability to see potential in everyone, we can have moments of proximity to God. *We could study how people of great love are able to gain humble insights in the nature of the universe.*

By the way, I still like your book on the story of a clam, still being a clammer at heart myself. Having dug tens of thousands of clams, I agree with you – we human beings now know as much about the nature of the universe and ultimate reality as a clam does about the nature of the sea.

I look forward to your comments. *It is my belief that the future of the Institute, at least for the next several years, could very successfully be focused on these five ideas alone.*

Gratefully,

Stephen Post

Cc: Dr. John M. Templeton, Jr., 610-825-2962
    Dr. Charles L. Harper, Jr. 610-825-1730
    Ms. Judith Marchand 610-825-1730
    Ms. Pamela Thompson 610-825-1730
    Dr. Gail Zummerman 307-266-3846

P.S. Hard copy in the mail to Lyford

To: ftbmw@coralwave.com
From: "Stephen G. Post" <sgp2@cwru.edu>
Subject: Mary - a special little favor, but Sir John may not wish to do this, and I will surely fully understand
Cc:
Bcc:
Attached: C:\Stephen's Stuff\Mike's Scale\SCALE Short 40 Only.doc;

Dear Mary:

Greetings. Hope all is well for you in 2006. I sense a great year for the world!
Irish intuition, that's all.

Mary, I am finishing up a major book on unlimited love, giving, and health/longevity for Random House. I have been working on it for two years. It is potentially very influential.

In the last chapter, which is in slow progress, I want to have a small section on a scale that measures people's unlimited love in various expressions. And, I am hoping to have a few notably generous and great philanthropists take the scale so we can briefly mention their character strengths (only). Do you think that Sir John would consider taking a few minutes to fill out the scale I have attached?

If so, then fax back to me at 216-368-8713. Or if you prefer, mail it back to me at

Stephen Post
Institute for Research on Unlimited Love
School of Medicine
Case Western Reserve University
Cleveland OH 44106-4976.

But I will fully appreciate Sir John's not wanting to bother with this. It would, however, be great if he could. I would be absolutely thrilled, and it would sure help get the message out.

Thanks a lot.

Cheers

Stephen

John M. Templeton
Box N-7776, Lyford Cay, Nassau, Bahamas
Telephone: (242) 362-4904 · Fax: (242) 362-4880

January 26th, 2006

Mr. Stephen Post
The Institute for Research on
Unlimited Love
Room 214, School of Medicine
Case Western Reserve University
10900 Euclid Avenue
Cleveland, Ohio 44106-8713

Dear Stephen:

Thank you for your e-mail. I have attached the text you sent down to me initialed by Sir John.

If there are any corrections, please let me know.

Sincerely yours,

Mary Walker

enc

SIR JOHN'S QUESTIONNAIRE RESULTS (S&P)

SIR JOHN

The Love Scale—Short Form
(The item numbers below refer to each items' original position on the development questionnaire)

Using the scale provided, please circle the one number that best reflects your opinion about whether or not each statement below describes you or experiences that you have had. There are no correct answers, so please respond as honestly as possible to each one.

1. I make a point of letting my family members know how much I appreciate them.
1=strongly disagree   2=disagree   3=slightly disagree   4=slightly agree   (5=agree)   6=strongly agree

15. I would not be where I am in life if it were not for the support of my friends.
1=strongly disagree   2=disagree   3=slightly disagree   (4=slightly agree)   5=agree   6=strongly agree

25. I see many things that people in my community do for which I am appreciative.
1=strongly disagree   2=disagree   3=slightly disagree   4=slightly agree   (5=agree)   6=strongly agree

32. When I hear about someone who has helped others, I feel appreciative that such people exist in the world.
1=strongly disagree   2=disagree   3=slightly disagree   4=slightly agree   (5=agree)   6=strongly agree

44. My loved ones know that if they have concerns, they can come to me and I'll give them the attention they need.
1=strongly disagree   2=disagree   3=slightly disagree   4=slightly agree   (5=agree)   6=strongly agree

53. When one of my friends needs my attention, I try to slow down and give them the time they need.
1=strongly disagree   2=disagree   3=slightly disagree   (4=slightly agree)   5=agree   6=strongly agree

62. My neighbors and co-workers know that they can come to me if they need to share their feelings with someone.
1=strongly disagree   2=disagree   (3=slightly disagree)   4=slightly agree   5=agree   6=strongly agree

74. I try to really pay attention to problems that are going on in the world.
1=strongly disagree   (2=disagree)   3=slightly disagree   4=slightly agree   5=agree   6=strongly agree

83. I can't resist reaching out to help when one of my family members seems to be hurting or suffering.
1=strongly disagree   (2=disagree)   3=slightly disagree   4=slightly agree   5=agree   6=strongly agree

93. I drop everything to care for my friends when they are feeling sad, in pain, or lonely.
1=strongly disagree   (2=disagree)   3=slightly disagree   4=slightly agree   5=agree   6=strongly agree

108. When people in my neighborhood or place of work are having problems, I do all I can to help them.
1=strongly disagree   (2=disagree)   3=slightly disagree   4=slightly agree   5=agree   6=strongly agree

118. I do not hesitate to lend my support to causes around the world that seek to help people who are unfortunate.
1=strongly disagree   (2=disagree)   3=slightly disagree   4=slightly agree   5=agree   6=strongly agree

128. I always go out of my way to help members of my family.
1=strongly disagree   2=disagree   (3=slightly disagree)   4=slightly agree   5=agree   6=strongly agree

138. It's personally important for me to be helpful to friends.
1=strongly disagree   2=disagree   (3=slightly disagree)   4=slightly agree   5=agree   6=strongly agree

149. If a neighbor or co-worker needs help, I offer it.
1=strongly disagree   (2=disagree)   3=slightly disagree   4=slightly agree   5=agree   6=strongly agree

1

SIR JOHN

160. I think it's important for me to try to leave this world better than I found it.
1=strongly disagree   2=disagree   3=slightly disagree   4=slightly agree   5=agree   6=strongly agree

171. My family can always count on me as if I were a "teammate."
1=strongly disagree   2=disagree   3=slightly disagree   4=slightly agree   5=agree   6=strongly agree

180. Friends know that they can always depend on me, rain or shine.
1=strongly disagree   2=disagree   3=slightly disagree   4=slightly agree   5=agree   6=strongly agree

189. I am more concerned about how I could help a co-worker or neighbor than about how much effort it could cost me.
1=strongly disagree   2=disagree   3=slightly disagree   4=slightly agree   5=agree   6=strongly agree

199. I am more concerned about how I could benefit society than about how much effort the activity could cost me.
1=strongly disagree   2=disagree   3=slightly disagree   4=slightly agree   5=agree   6=strongly agree

208. I believe that I always can gain something from considering the perspectives of my family members.
1=strongly disagree   2=disagree   3=slightly disagree   4=slightly agree   5=agree   6=strongly agree

221. Regardless of a friend's origins, upbringing, or background, I try to communicate my respect for them.
1=strongly disagree   2=disagree   3=slightly disagree   4=slightly agree   5=agree   6=strongly agree

228. I believe that I always can gain something from hearing the perspective of my co-workers or neighbors.
1=strongly disagree   2=disagree   3=slightly disagree   4=slightly agree   5=agree   6=strongly agree

240. People should make it a point to acknowledge the efforts and aspirations of others around them.
1=strongly disagree   2=disagree   3=slightly disagree   4=slightly agree   5=agree   6=strongly agree

256. I'm not good at helping family members figure out their strengths in life.®
1=strongly disagree   2=disagree   3=slightly disagree   4=slightly agree   5=agree   6=strongly agree

258. I get pleasure out of using my creative skills to help friends with valuable projects.
1=strongly disagree   2=disagree   3=slightly disagree   4=slightly agree   5=agree   6=strongly agree

267. Time does not matter when it comes to helping neighbors or people at work develop a creative idea.
1=strongly disagree   2=disagree   3=slightly disagree   4=slightly agree   5=agree   6=strongly agree

283. Coming up with ways to create opportunities for other people doesn't interest me.®
1=strongly disagree   2=disagree   3=slightly disagree   4=slightly agree   5=agree   6=strongly agree

288. I use humor to try to give my family a fresh perspective and hope.
1=strongly disagree   2=disagree   3=slightly disagree   4=slightly agree   5=agree   6=strongly agree

301. Sharing funny experiences with friends is uplifting.
1=strongly disagree   2=disagree   3=slightly disagree   4=slightly agree   5=agree   6=strongly agree

312. I don't try to make neighbors or co-workers laugh when they are stressed out.®
1=strongly disagree   2=disagree   3=slightly disagree   4=slightly agree   5=agree   6=strongly agree

319. I usually try to break the ice or improve the atmosphere with humor or comical stories.
1=strongly disagree   2=disagree   3=slightly disagree   4=slightly agree   5=agree   6=strongly agree

328. I am willing to confront my family members when they do something that is harmful to others.

2

1=strongly disagree   2=disagree   3=slightly disagree   4=slightly agree   5=agree   6=strongly agree

344. I'm too timid to confront my friends when their behavior is hurting themselves or another person.®
1=strongly disagree   2=disagree   3=slightly disagree   4=slightly agree   5=agree   6=strongly agree

351. I am willing to take personal risks in my neighborhood or place of work to insure that everyone is treated fairly.
1=strongly disagree   2=disagree   3=slightly disagree   4=slightly agree   5=agree   6=strongly agree

357. I have supported social organizations that are devoted to correcting injustices in the world.
1=strongly disagree   2=disagree   3=slightly disagree   4=slightly agree   5=agree   6=strongly agree

374. It is not very easy to forgive when a family member hurts me.®
1=strongly disagree   2=disagree   3=slightly disagree   4=slightly agree   5=agree   6=strongly agree

377. I never hang onto grudges when one of my friends does something that hurts me.
1=strongly disagree   2=disagree   3=slightly disagree   4=slightly agree   5=agree   6=strongly agree

391. I try to set an example of forgiveness in my community and place of work.
1=strongly disagree   2=disagree   3=slightly disagree   4=slightly agree   5=agree   6=strongly agree

401. Forgiveness should be a much bigger part of foreign relations.
1=strongly disagree   2=disagree   3=slightly disagree   4=slightly agree   5=agree   6=strongly agree

3

SIR JOHN

1.		312.	2 − R = 5
15.	4	319	5
25.	5	328	2
32	5	344	2 − R = 5
44.	5	351	5
53.	4	357	6
62.	3	374	1 − ℝ = 6
74.	2	377	6
83.	2	391	5
93.	2	401	5
108.			
118.	2		
128.	3	157	
138	3		
149.	2	! 20th percentile	
160.	6		
171.	4		
180.	4		
189.	3		
199	6		
208	4		
221	5		
228	4		
240	3		
256	2 − ℝ = 5		
258	3		
267	3 − ℝ = 4		
283	3		
288	3		
301	5		

doesn't
drop everything
to help the
suffering with
untold distress,
but works out
to leave the world
a better place 6

To: ftbmw@coralwave.com
From: "Stephen G. Post" <sgp2@cwru.edu>
Subject: Thanks to Sir John, and one more little favor
Cc:
Bcc:
Attached:

Dear Mary:

Many thanks to Sir John for faxing me the love scale. It is terrific to have it.

Random House is the world's largest publisher, and they are featuring my book on love in the non-fiction category in the fall of 2006, or possibly very early 2007. They expect a best seller. Of course, I mention Sir John as the most generative and inspiring and visionary man I have had the honor of knowing. This is the great book that will bring all the science and practice and story and "spirituality" of love together for the world. I had to do this, and am now on the last chapter.

Random House gave a major advance, which I have used to support writers helping me in three different cities.

This has been the most difficult writing project of my life, but also the very best by far. Sir John will be so proud!

I am wondering, Mary, if Sir John could write just a paragraph or so, of even just send me a few fresh lines for the book, on why love is the center of all virtue, why it is the only credible energy for the human future, and how he felt taking the little scale, and what his hopes are for the future of "unlimited love" as a field.

This could just be a few lines. But I am dedicating this work to Sir John, who has always treated me and all those around him with love.

If he could do this, please fax to me at 216-368-8713. I want to place it in the very inspiring last chapter.

Thanks. And please let Sir John know that, after 5 years of full time dedication to the "unlimited love" project, what I discover is that "Love begets love." He is an example of that.

All Gratitude

Stephen

John M. Templeton

Box N-7776, Lyford Cay, Nassau, Bahamas

Telephone: (242) 362-4904 · Fax: (242) 362-4880

"Love is the most powerful dynamic in our lives and in the universe, and the more we learn about it and practice it, the better off the future will be. We should love everyone without exception, but that doesn't mean we condone what they do or say. And sometimes, in practicing the love of neighbor well, we have to be very careful to encourage them to become responsible for themselves. As a philanthropist, "love of humanity" can mean finding ways of encouraging thrift, gratitude, responsibility, and character and so that people can discover their creative gifts and use them for noble purposes. Love is helping people to grow in creativity and discovery and vision. Sometimes the worst thing you can do for someone is to solve problems that they can and ought to be solving for themselves. I believe that God is love, and that the more we can discover that "love begets love" and find out for ourselves how this energy of love can enhance our lives, the better off we will be. Creativity, vision, freedom, individual responsibility – these are laws of life. Love is the power that binds the universe, that harmonizes, that creates without destroying, and then enhances every aspect of a persons life. Taking the scale of love let me reflect on love in all its various manifestations. And love blesses not just those who receive, but perhaps even more, those who give. That is what I have lived by."

Acknowledgments

I AM GRATEFUL TO Sir John Templeton (1912–2008) for teaching me to investigate sublime spiritual love not only as reported by individuals who claim to have directly experienced it, but additionally as a matrix or substrate underlying and sustaining all of reality. What a strange idea to us modernists for whom the universe has become disenchanted and demystified. Yet great spiritual minds from every religious tradition have asserted that beneath the ordinary surface perception of visible reality there is an Ultimate Reality, or what Paul Tillich called a "Ground of Being." Sir John wondered if in the future science and spiritualty might converge on the same perennial truth with their very different ways of knowing. I thank Sir John for being humbly audacious in thinking big, for encouraging us to do so with him, and for creating the title of this book, *Is Ultimate Reality Unlimited Love?* Modern consciousness, desperately unable to find adequate meaning in sensate materialism, is turning in a spiritual direction that makes Sir John's big question both delightfully refreshing and hopeful.

Sir John has a loyal, diligent, and kind son. I thank Dr. John M. Templeton Jr., MD, for informing me of his father's request, made during Sir John's final days on earth, that a book to be written to consolidate faithfully his lifetime of reflection on the two conjoined ideas that meant so much to him—Ultimate Reality and Unlimited Love. It was Dr. Templeton who conveyed to me the invitation to write. I thank Dr. Templeton's beloved wife, Dr. Josephine "Pina" Templeton, MD, for her reliable inspiration as this book took shape. Equally, I thank Barnaby Marsh of the John Templeton Foundation for his valued friendship over the years. Barnaby too knew Sir John well and grasps how much these two ideas meant to him.

Thanks to Susan Arellano, editor-in-chief of the Templeton Press. A splendid editor, Susan had to ask me some hard questions. How can a book about Ultimate Reality and Unlimited Love connect with the modern reader? How could it be convincing? What did Sir John mean by all this? I responded that this book is primarily intended for the talented individuals being recruited now and in the future to work with the Templeton Foundation, some of whom may be unaware of—or even shy away from—Sir John's very significant metaphysical interests. After all, these interests are so deeply contrary to the paradigms of the modern university researcher—paradigms that Sir John hoped to see change. Moreover, I responded that this book is less about answering Sir John's big question with incontrovertible proof than it was about indicating plausible lines of evidence for future investigations even many years from today. Susan, to her credit, recognized that this book had to be written if we are to take Sir John seriously in the future. A special note of appreciation is due to one of Susan's right hands at the press, Trish Vergilio. As we reached the final stages of editing and production, Trish and I teamed up, and that synergy resulted in a better book.

I thank my loyal and beloved Cleveland friends Cathy M. Lewis, Kathy Pender, Judith B. Watson, Joni Marra, Matthew T. Lee, and Richard T. Watson (1933–2011), all members of the Board of Directors of the Institute for Research on Unlimited Love, which was founded with Sir John's support in 2001 (www.unlimitedloveinstitute.com). These are people of special courage, spiritual vision, and immense generosity.

I also thank so many good people from Greater Cleveland who remained my friends after I left in 2008, having served for twenty years at the School of Medicine of Case Western Reserve University. Among those friends was the beloved neurologist Joseph Michael Foley, MD (1916–2012), a Roman Catholic who never believed that mind and spirit could be fully explained by brain matter.

The Rev. William Eddy (1924–1989) of Christ Church Tarrytown deserves a note of gratitude because he was so influential in my renewed personal affirmation of basic Christianity in 1985. We had many wonderful discussions of physics and theology in his study during those

Tarrytown years. So also my thanks to the Rev. Richard Neuhaus (1936–2009), who invited me into his small group for conversations about the relationship between faith and science at the Union Club in Manhattan before I left Tarrytown for Cleveland in 1988. Thanks to fellow conversationalists Sydney Callahan, Peter Berger, Stanley Hauerwas, Christopher Lasch (1932–1994), and Gilbert Meilaender.

Much appreciation is due to James M. Gustafson, Don S. Browning (1934–2010), Langdon Gilkey (1919–2004), Martin E. Marty, Robin W. Lovin, David Tracy, Joseph Kitagawa, and others who were important influences on my thinking about science and religion in my formative years as a student at the Divinity School of the University of Chicago.

Humble thanks to my wife Mitsuko, daughter Emma, and son Drew for their support as I have pursued spiritual progress in Unlimited Love with passion for so many years, knowing that such pursuits can be hard for any family to fathom.

Lastly, let me state that I did not write a word of this book at my campus office in the Department of Preventive Medicine at Stony Brook University, where I work ten hours a day entirely focused on educating students in medical humanities, compassionate care, and medical ethics. This book was written at home on Sundays and in the earliest hours of the morning when the earth is still quiet and pristine. However, I am grateful to the many good people around Stony Brook and Greater New York who are friends and supporters.

Stephen G. Post, PhD
President, The Institute for Research on Unlimited Love
(www.unlimitedloveinstitute.com)
Trustee of the Templeton Foundation (2008–2014)
Thanksgiving 2013

Bibliography

A.A. World Services. 1939/2001. *Alcoholics Anonymous*. Fourth ed. New York: American Book-Stratford Press.

Albert, David. 2012. "On the Origin of Everything." *New York Times Book Review*. Review of Krauss, *Universe from Nothing*. March 23.

Allen, James. 1903/2008. *As a Man Thinketh*. New York: Jermey P. Tarcher/Penguin.

Auden, W. H. 1965. "Introduction." In *The Protestant Mystics: An Anthology of Spiritual Experience from Martin Luther to T. S. Eliot*. Edited by Anne Fremantle, 3-43. New York: Mentor Books.

Bergson, Henri. 1932/1935. *The Two Sources of Morality and Religion*. Trans. R. Ashley Audra and Cloudesley Brereton. Garden City, NY: Doubleday & Co. Original in French.

Boyer, Pascal. 2001. *Religion Explained*. New York: Basic Books.

Brooks, David. 2010. "Bill Wilson's Gospel." *New York Times*.

Brunner, Emil. 1932/1948. *The Divine Imperative*. Second English ed. Philadelphia: Westminster Press.

Cady, H. Emilie. 1900–1930/1995. *Complete Works of H. Emilie Cady*. Unity Village, MO: Unity Books.

Chittick, William C. 2010. "Divine and Human Love in Islam." In *Divine Love: Perspectives From the World's Religious Traditions*, ed. Jeffrey Levin and Stephen G. Post, with a Foreword by Seyyed Hossein Nasr, 163–200. Philadelphia: Templeton Press.

Davies, Paul. 1992. *The Mind of God: The Scientific Basis for a Rational World*. New York: Simon & Schuster.

Dawkins, Richard. 2006. *The God Delusion*. Boston: Houghton Mifflin.

Dixon, Thomas. 2008. *The Invention of Altruism: Making Moral Meanings in Victorian Britain*. New York: Oxford University Press.

Durkheim, Emil. 1912/1995. *The Elementary Forms of Religious Life.* Trans. Karen E. Fields. New York: Free Press.

Dyson, Freeman J. 1997. *Imagined Worlds.* Cambridge, MA: Harvard University Press.

Ellis, George F. R. 1993. The Theology of the Anthropic Principle." In *Quantum Cosmology and the Laws of Nature,* ed. R. J. Russell, N. Murphy, and C. J. Isham, 367–406. Vatican Observatory/CTNS.

————. 2006. "Issues in the Philosophy of Cosmology." In *Handbook in Philosophy of Physics,* ed. J. Butterfield and J. Earman, 183–285. New York: Elsevier.

————. 2008. "Faith, Hope and Doubt in Time of Uncertainty." The James Backhouse Lecture: Australia Yearly Meeting of Quakers.

————. 2011. "Why Are the Laws of Nature as They Are? What Underlies Their Existence?" In *The Astronomy Revolution: 400 Years of Exploring the Cosmos,* ed. Donald York, Owen Gingerich, and Shuang-Nan Zhang,. London/Boca Raton: CRC Press/Taylor & Francis Group.

Ellis, George F. R., D. Noble, and T. O'Connor. 2012. "Top-Down Causation: An Integrating Theme across the Sciences?" *Interface Focus* 2.

Fillmore, Charles. 1999. *The Essential Charles Fillmore: Collected Writings of a Missouri Mystic,* ed. with commentary by James Gaither. Unity Village, MO: Unity Books.

Ford, Kenneth W. 2004. *The Quantum World: Quantum Physics for Everyone.* Cambridge, MA: Harvard University Press.

Gaita, Raimond. 2004. *Good and Evil: An Absolute Conception.* New York: Routledge.

Gandhi, Mahatma K. 1957/1970. *The Law of Love.* Bombay: Bharatiya Vidya Bhavan.

Gould, Stephen Jay. 2002. *Rock of Ages: Science and Religion in the Fullness of Life.* New York: Ballentine Books.

Hallie, Philip. 1994. *Lest Innocent Blood Be Shed: The Story of the Village of Le Chambon and How Goodness Happened There.* New York: Harper Perennial.

Hawking, Stephen. 1988. *A Brief History of Time: From the Big Bang to Black Holes.* New York: Bantam Books.

Hawking, Stephen, and Leonard Mlodinow. 2010. *The Grand Design.* New York: Bantam.

Henry, Richard Conn. 2005. "The Mental Universe." *Nature* 436, no. 7. July.

Herrmann, Robert L. 2004. *Sir John Templeton: Supporting Scientific Research for Spiritual Discoveries.* Philadelphia: Templeton Foundation Press.

Hill, Napoleon. 1937/2005. *Think and Grow Rich.* New York: Jeremy P. Tarcher/Penguin.

Holmes, Ernest. 1938/1998. *The Science of Mind: A Philosophy, A Faith, A Way of Life.* New York: Jeremy P. Tarcher/Putnam.

Huxley, Aldous. 1945. *The Perennial Philosophy.* New York: Harper Perennial.

James, William. 1902/1982. *The Varieties of Religious Experience.* New York: Penguin Books.

Jeans, James. 1943. *The Mysterious Universe.* Revised ed. New York: Macmillan. Kessinger Legacy Reprints.

Johnston, Barry V. 1995. *Pitirim A. Sorokin: An Intellectual Biography.* Lawrence: University Press of Kansas.

Kaufman, Gordon. 2000. *In the Beginning . . . Creativity.* Minneapolis: Augsberg Fortres.

Kierkegaard, Søren. 1847/1962. *Works of Love.* Trans. Howard and Edna Hong. New York: Harper & Row.

King, Ursula. 2004. "Love: A Higher Form of Human Energy in the Work of Teilhard de Chardin and Sorokin." *Zygon* 39, no. 1 (March): 77–102.

Kornberg Greenberg, Yudit, editor-in-chief. 2008. *Encyclopedia of Love in World Religions.* 2 vols. Santa Barbara, CA: ABC-CLIO.

Krauss, Lawrence M. 2012. *A Universe from Nothing: Why There is Something Rather than Nothing.* New York: Free Press.

Lee, Matthew T., and Margaret N. Poloma. 2009. *A Sociological Study of the Great Commandment: The Practice of Godly Love in Benevolent Service.* Lewiston, NY: Edwin Mellen Press.

Lee, Matthew T., Margaret M. Poloma, and Stephen G. Post. 2013. *The Heart of Religion*. New York: Oxford University Press.

Levin, Jeffery R., and Stephen G. Post, eds., 2010. *Divine Love: Perspectives from the World's Religious Traditions*. Philadelphia: Templeton Press.

Marmot, Michael. 2005. *The Status Syndrome: How Social Status Affects Health and Longevity*. New York: Holt.

Milosz, Czaslaw. 2003. *New and Collected Poems, 1931–2001*. New York: Harper Collins.

Mosley, Glenn R. 2006. *New Thought, Ancient Wisdom: The History and Future of the New Thought Movement*. Philadelphia: Templeton Foundation Press.

Murphy, Nancey, and George F. R. Ellis. 1996. *On the Moral Nature of the Universe: Cosmology, Theology, and Ethics*. Minneapolis: Fortress Press.

Nagel, Thomas. 2012. *Mind & Cosmos: Why the Materialist Neo-Darwinian Conception of Nature is Almost Certainly Wrong*. New York: Oxford University Press.

Newberg, A., E. G. D'Aquili, and L. Rause. 2001. *Why God Won't Go Away: Brain Science and the Biology of Belief*. New York: Ballantine.

Pagano, M. E., K. B. Friend, J. S. Tonigan, and R. L. Stout. 2004. "Helping Other Alcoholics in Alcoholics Anonymous and Drinking Outcomes: Findings from Project MATCH." *Journal of Studies on Alcohol* 65, no. 6: 66–73.

Pagano, M. E., B. Zeltner, J. Jaber, W. H. Zywiak, and R. L. Stout, 2009. "Helping Others and Long-Term Sobriety: Who Should I Help to Stay Sober?" *Alcohol Treatment Quarterly* 27, no. 1: 38–50.

Peale, Norman Vincent. 1952. *The Power of Positive Thinking*. New York: Fawcett Crest.

Pierce, Charles S. 1893/1955. "Evolutionary Love." In *Philosophical Writings of Pierce*, ed. Justice Buchler, 361–74. New York: Dover.

Pigliucci, Massimo. 2012. "Lawrence Krauss: Another Physicist with an Anti-Philosophy Complex." *Rationally Speaking* (blog). April 25.

Post, Stephen G., 2012. "The Ontological Generality in Spirituality and Health." In *Healing to All Their Flesh: Essays in Spirituality, Theology, and Health*, ed. Jeff Levin and Keith G. Meador. 186–218. West Conshohocken: Templeton Press.

Post, Stephen G., Lynn G. Underwood, Jeff R. Schloss, and William B. Hurlbut. 2002. *Altruism and Altruistic Love: Science, Philosophy and Religion in Dialogue*. New York: Oxford University Press.

Post, Stephen G., ed. 2007. *Altruism and Health: Perspectives from Empirical Research*. New York: Oxford University Press.

Proctor, William. 1983. *The Templeton Touch*. Garden City: Doubleday and Company, Inc.

Rees, Martin. 2000. *Just Six Numbers: The Deep Forces That Shape the Universe*. New York: Basic Books.

Rose, Stuart. 2007. *Sublime Love: Essay and Anthology*. Indica Books: Varanasi, India.

Scherwitz, L., R. McKelwain, C. Laman, J. Patterson, L. Dutton, S. Yusim, et al. 1983. "Type A Behavior, Self-Involvement, and Coronary Atherosclerosis." *Psychosomatic Medicine* 45, no. 1: 47–57.

Smith, Huston. 2001. *Why Religion Matters: The Fate of the Human Spirit in an Age of Disbelief*. San Francisco: Harper San Francisco.

Sorokin, Pitirim A. 1937. *Social and Cultural Dynamics*, 3 vols. New York: American Book Co.

———. 1954/2002. *The Ways and Power of Love: Types, Factors, and Techniques of Moral Transformation*. With an Introduction by S. G. Post. Philadelphia: Templeton Press.

Stenger, V. J. 2007. *God, the Failed Hypothesis: How Science Shows That God Does Not Exist*. New York: Prometheus Books.

Susskind, Leonard. 2005. *The Cosmic Landscape: String Theory and the Illusion of Intelligent Design*. Boston: Little, Brown, & Co.

Temple, William. 1985. *Readings in St. John's Gospel*. New York: Morehouse Barlow.

Templeton, John Marks. 1981/1995. *The Humble Approach: Scientists Discover God. Can Science Expand Our Understanding of the Divine?* Philadelphia: Templeton Foundation Press.

Templeton, John Marks (as described to James Ellison). 1987. *The Templeton Plan: A Tried and True Guide to Personal Growth and Well-Being*. New York: Harper Paperbacks.

Templeton, John Marks. 1997. *Worldwide Laws of Life: 200 Eternal Spiritual Principles.* Philadelphia: Templeton Foundation Press.

———. 1999. *Agape Love: A Tradition Found in Eight World Religions.* Philadelphia: Templeton Foundation Press.

———. 2000a. *Possibilities for Over One Hundredfold More Spiritual Information: The Humble Approach in Theology and Science.* Philadelphia: Templeton Foundation Press.

———. 2000b. *Pure Unlimited Love: An Eternal Creative Force and Blessing Taught by All Religions.* Philadelphia: Templeton Foundation Press.

———. 2001. *Story of a Clam: A Fable of Discovery and Enlightenment.* Philadelphia: Templeton Foundation Press.

Templeton, John Marks (with Rebekah Alezander Dunlap). 2003. *Why Are We Created? Increasing Our Understanding of Humanity's Purpose on Earth.* Philadelphia: Templeton Foundation Press.

Templeton, John Marks. 2012. *The Essential Worldwide Laws of Life.* With a Foreword by Stephen G. Post. West Conshohocken: Templeton Press.

Templeton, John Marks, and Robert L. Herrmann. 1989/1998. *The God Who Would Be Known: Revelations of the Divine in Contemporary Science.* Philadelphia: Templeton Foundation Press.

———. 1994. *Is God the Only Reality? Science Points to a Deeper Meaning of the Universe.* New York: Continuum.

Templeton, John Marks, ed., and with various of his own entries, 1990/2006. *Riches for the Mind and Spirit: John Marks Templeton's Treasure of Words to Help, Inspire & Live By.* Philadelphia: Templeton Foundation Press.

Templeton, John Marks, ed. 2000. *Worldwide Worship: Prayers, Songs, and Poetry.* Philadelphia: Templeton Foundation Press.

Wilber, Ken. 2001. *Quantum Questions: Mystical Writings of the World's Greatest Physicists.* Boston: Shambhala.

Index